The Science of the Oven

Arts and Traditions of the Table

ARTS AND TRADITIONS OF THE TABLE

Albert Sonnenfeld, Series Editor

The Science of the Oven

Hervé This

Translated by Jody Gladding

Columbia University Press *New York*

Columbia University Press

Publishers Since 1893

New York   Chichester, West Sussex

Copyright © Pour la science 2007

Translation copyright © 2009 Columbia University Press

All rights reserved

Library of Congress Cataloging-in-Publication Data

This, Hervé

[De la science aux fourneaux. English]

The science of the oven / Hervé This ; translated by Jody Gladding.

p. cm.

Includes bibliographical references and index.

ISBN 978-0-231-14706-4 (cloth : alk. paper) — ISBN 978-0-231-51854-3 (ebook)

1. Food — Sensory evaluation. 2. Food — Analysis. 3. Food — Composition.

4. Gastronomy. I. Title.

TX546.T55313   2009

641.01'3 — dc22   2009007523

Columbia University Press books are printed on permanent and durable acid-free paper.

This book is printed on paper with recycled content.

Printed in the United States of America

c 10 9 8 7 6 5 4 3 2 1

The impossibility of isolating nomenclature from science and science from nomenclature stems from the fact that all physical science is necessarily based on three things: the series of facts that constitute the science, the ideas that recall them, the words that express them. . . . Since it is words that preserve ideas, and transmit them, the result is that languages cannot be perfected without perfecting science, nor science without language.

Antoine Laurant de Lavoisier, *Traité de chimie*

Contents

The Science of the Oven

What is cooking? And why do so many of us enjoy it? Basically, if cooking consists of roasting meat, tenderizing vegetables in water, binding sauces, kneading dough . . . there is not really much to making . . . a dish.

Roasting meat? The steps are dishearteningly banal: You take a piece of meat, put it on a spit, heat it, take it off the spit. Many of those who cook on a daily basis grow weary of this "chore" that an allocation of domestic tasks assigns them.

Tenderizing vegetables by cooking them "English style"? The term hardly hides the intellectual poverty of the work: Just take a saucepan, fill it with water, maybe add a handful of coarse salt (what a noble gesture!), heat it (in the past, there was the uncertainty of lighting a fire, but today, the electric burner comes on every time), then put in the vegetables and wait. Can we elevate our minds through such basic gestures? Even those devoted to the Zen art of tea or archery will find it hard to persuade themselves that dropping vegetables into a saucepan requires an infinitely moral asceticism!

Binding a sauce? No need for a Harvard education to do that. To the juices you want to thicken, you can either add a few spoonfuls of flour or whisk in fat or some sort of protein (found in an egg, blood, etc.) generally without knowing you are doing so, and heat gently.

Kneading? That is not difficult either; you stretch, fold, stretch, fold, stretch, fold. . . .

So why cook, except to fill the stomach and simply survive? Because, if none of the aforementioned techniques is difficult, all are mysterious when you stop to examine their effects. For example, when mayonnaise "takes," it is a matter of liquids (egg yolk, vinegar, oil) that form a semisolid consistency; when an egg coagulates, it is a liquid that hardens when heated, whereas solids melt when heated; when meat is roasted, its surface browns and acquires flavor. . . . What accounts for these transformations, which are extraordinary in the most literal sense of the term?

Yes, the boredom of cooking only affects those who casually pass over such phenomena without seeing them, those who limit themselves to techniques without paying attention to the results. And so. . . . The act of cooking is boring when it is solely technique, without technology, science, or art.

IMPOSSIBLE TECHNOLOGY!

Technique is doing it, and the very existence of those who are bored by cooking proves that doing it can be the execution of the gesture alone, with no thought to the way the gesture is done. That said, why is it that those bored cooks do not take advantage of the technological potential technique offers them? The question calls for a specific response to cooking, the last "chemical art" left unsystematized until the creation of its scientific discipline, "molecular gastronomy."

Why do we still cook as we did in the Middle Ages, with whisks, fire, and saucepans? Why this outdated behavior, when, at the same time, humanity is sending probes to the outer limits of the solar system? Why do our recipes hardly differ from those found in the *Viandier*, a treatise by Guillaume Tirel, known as Taillevent, who lived in the fourteenth century and whose recipes, moreover, hardly differ from those in Apicius's *De re coquinaria*, a collection of texts assembled between the fourth and fifth century AD? Why this noticeable technical stagnation?

Let us look at culinary transformations from the perspective of cooks twenty-five years ago. We have forgotten that when I offered them the use of carrageenans for jelling liquids, ultrasonic sound tanks for emulsifying fats, rotary evaporators for reducing bouillons, the questions that almost invariably arose concerned the safety of my proposals.

It is a truism to state that dishes are made to be consumed and that we cannot eat with impunity every kind of matter whatever, animal, vegetable, or a mixture of the two. It took thousands of years for humanity to learn to recognize (and indeed we are still learning) which plants can be safely consumed, which parts of animals are edible. Cookbooks from the past show that such knowledge is still "fresh." Until the twentieth century, mushrooms were identified by bringing them into contact with iron; the poisonous ones were supposed to turn black! Since the Renaissance, pike eggs have been considered toxic, but who is willing to take the risk of verifying this?

Human lives were lost to determine which foods were comestible and which foods were to be avoided. Moreover, the knowledge thus wrested from death, so to speak, is unreliable; smoked products, for example, are loaded with carcinogenic molecules that cause digestive cancers, and many accepted food products warrant reconsideration. Interactions between foods are suspect as well. Humans limited themselves to a small repertoire for fear of dying from what they ate; they were endowed with prudence, which proved their salvation, which is why we retain our medieval behavior with regard to food. And the result is technical stagnation; since to eat anything new is to expose oneself to danger, conservatism is a sound approach and introducing new technology in cooking is next to impossible.

SCIENCE AND TECHNOLOGY

What is more, technology deserves to be broken down a bit. I am distinguishing a "local" technology, which is limited to the perfecting of old techniques, to the fine-tuning proposed by one with a knack for improvements, from a "global" technology that makes use of new knowledge contributed by science.

Local technology would consist of understanding that the whisk that introduces air bubbles into a stiffly beaten egg white is more effective when it has more wires; consequently, such technology suggests multiplying the wires in a whisk. As for global technology, it calls into question why a whisk is used for beating egg whites. Why not use, instead, a compressor and a nozzle that can introduce air bubbles into the egg white? Or any other entirely different device that performs the same operation of increasing mass more effectively than a whisk? That is what the "pianocktail" does (see "Making the Pianocktail" in chap. 7).

To establish the distinction between local and global technology, let us add that the pianocktail owes only its name to the French writer Boris Vian (1920–1959); it was conceived while we were seeking a practical application for a system we had perfected for describing dispersed complex systems. Nothing to do with cooking!

Moreover, science has nothing to do with applications. It has another objective: the production of knowledge! The expression "applied science" is a serious intellectual offense, and Louis Pasteur rose up against it many times: "An essentially false idea has entered into numerous discussions arising from the creation of a secondary professional education, which is that applied sciences exist. There are no applied sciences. The very combination of those words is shocking. But there are applications of science, which is very different." Still worse, Pasteur nearly explains the defeat of 1870 by the confusion between science and technology: "No, a thousand times no, there does not exist a category of science to which the name of applied sciences can be given. There is science and the applications of science, bound together as the fruit is bound to the tree that bore it."

In the same vein, we should avoid speaking of "basic sciences," since knowledge has no boundaries and since attributing it to a particular discipline is a way of demarcating a territory (in order to retain all the credit?) or exempting oneself (out of laziness?) from acquiring the knowledge neighboring territories offer. We should also avoid lapsing into August Comte's great mistake, which was wanting to hierarchize the sciences!

AND THE PLACE OF COOKING IN ALL THIS?

Let us return to the kitchen. The phenomena observed there have their science, molecular gastronomy, which I created in 1988 with the British physicist Nicholas Kurti. In retrospect, it is necessary to admit that if the idea was clear, the initial program was faulty. Besides, cooking had been the object of studies in the past, had it not? As described on an Egyptian tablet, the experiment of weighing fermented meat to learn if it lost an "emanation" was already scientific, since it involved the search for a mechanism to explain a phenomenon.

Of course the experimental method we owe to Bacon, Galileo, and Palissy was not as clearly established as it is today, and arithmetic—let us distinguish this from mathematics, the term that designates the entire discipline—had not yet been recognized as the theoretical safeguard it has become. Nevertheless it was already a matter of studying culinary phenomena, and that experiment of weighing fermenting meat is part of the prehistory of molecular gastronomy. As are the experiments of the great Antoine Laurent de Lavoisier on the "suitable" proportions of meat and water for making bouillon. Or the studies of Justus von Liebig, Alexis Cadet de Vaux, Michel-Eugène Chevreul. . . .

In the past, however, food science was not as fussy about details as it is today. Subtle characterizations of foods were mixed up with the study of culinary transformations and also with improving food industry processes. In 1988, molecular gastronomy took its rightful place between the science of food and the technology of processing. Today it is devoted to . . . to what, exactly?

To culinary technique—finally we come back to that! The techniques mentioned at the very beginning of the introduction and the phenomena that accompany them, during the roasting of meat, the kneading of dough, the binding of a sauce, are the objects of investigations . . . thus avoiding the boredom that comes from disinterest (and not from uniformity; the world hardly changes!). The study of this component of molecular gastronomy was the subject of the first works. While the results were accruing, the discipline extricated itself from its "original sin," the confusion with technology.

Yes, the discipline's program was flawed, since it entailed the following five objectives: (1) to explore recipes; (2) to collect and test sayings, cooking tricks, adages, proverbs, and so on; (3) to invent new dishes; (4) to introduce new tools, utensils, ingredients; and (5) to use the general appeal of cooking as a means for demonstrating the beauties of science in general, and chemistry in particular. Wrong! Objective 5 is political, or perhaps social. Objectives 3 and 4 are technological.

In 2003, molecular gastronomy's program was redefined while "molecular cuisine" was becoming the fashion throughout the world. After two decades of work, today's cooks use liquid nitrogen to make their ice cream and sorbet, and better still, they distill, infuse, and jell with the aid of jelling agents long used by the food industry (the transition is due especially to the mad cow crisis, which caused gelatin to be rejected for "new" jelling agents).

Simultaneously, it became apparent—why had no one seen it before?— that every recipe is composed of three parts, that is, one part that is technically useless, one part that gives the "definition" of the dish, and a part that gives the "detailed information." Under this last heading, let us group the techniques, tricks, adages, proverbs, hearsay, sayings. . . . For example, mayonnaise is obtained by dispersing oil in a mixture of egg yolk and vinegar; that is the definition. Adding lemon juice at the end of the preparation; that is a bit of "detailed information."

Better still, these definitions and details must be analyzed from the triple point of view of love, art, and technique. Making mayonnaise that is an oil-in-water type emulsion is the technical component. Making mayonnaise that has a "good" flavor is a question of art. Making mayonnaise that pleases the dinner guest is a question of love, in the sense of wanting to provide happiness to others.

AN ONGOING PROJECT: KNOWLEDGE

If the objectives are now clear, the work must be ongoing. Each month, in the journal *Pour la Science*, especially, we report scientific results and seek their culinary applications. Those results and culinary applications are gathered here.

Experience shows that the "provisions" for food happiness are threefold: an exploration of sensorial physiology, a knowledge of the effects of foods on the organism, and a knowledge of ingredients (what cooks call "products"). Sensorial neurophysiology "feeds" the first part, physiology and toxicology the second part, and agronomy the third part. Nevertheless, disciplinary fields are not strictly divided and disciplinary boundaries are not clear, or useful. The objective is always to understand, is it not?

Down with disciplinary boundaries! All that matters is the subject that captivates us: cooking! If quantum mechanics was key to a potential delicacy, would we not make every effort to manipulate operators in the Hilbert spaces? Is the gourmandise drive for knowledge such a bad thing?

In any case, what the first part of our provisions kit shows is that false ideas, as numerous as knowledge gaps, spread with regard to cooking and to tasting. If it is now clear to sensorial neurophysiologists that the number of tastes is not four, for example, books still contain "maps of the tongue" that claim that the tip recognizes sweet, and so on. Wrong! Wrong! Wrong! To witness that all tongues are different, all you have to do is ask a group of individuals to stick their tongues into sweet or salty water. Which raises the question: Why do we cling so to these false ideas on sensorial physiology?

Likewise, it is claimed—even by "specialists"—that smell constitutes 90 percent of taste, but this value has never been measured! Why do we swallow such nonsense? Is it because, with a cold, we sometimes lose all perception of taste? To those who might be tempted to take that as proof, let us remember that we also lose our sense of taste when we burn our mouths on excessively hot food.

No, our daily gustatory experiences do not make us good connoisseurs of gustatory physiology, and science plays that marvelous role of often refuting us. "Even so, I am not insane enough to be completely sure of my certainties," said the biologist Jean Rostand. It also plays the superb role of showing us what we cannot see, cannot even imagine, "of lifting a corner of the great veil," said Albert Einstein. It can even give rise, through illusions, to sensations that have no grounds for existing!

When ingested, foods produce effects. In these times of state-guaranteed citizen comfort, the term "food health and safety" is ubiquitous. We have forgotten that, first of all, a sufficient quantity of food to maintain the organism was the basic issue. That we are preoccupied with the flavor and quality of tomatoes in winter is striking proof of modern agronomy's success. As a result, science has been reduced to examining the details, losing sight of the bigger picture. It tracks the trace element, the antioxidant molecule that—we believe will help us avoid our all too inevitable aging, it explores the virtues of foods, and it ends up establishing sometimes very limited observations as dogma.

Of course, restricting oneself to an insufficiently varied or unbalanced diet is not healthy. All the same, the question of "healthy food" is not a new one, since it was already a concern for medieval cooking, where imbalances in temperament were countered with—false—ideas casually extrapolated from the—equally false—Aristotelian theory of the four elements: earth, water, air, fire.

Today, it is easy for the food industry to see in foods the particular virtues of their molecular components; the number of molecules is so great, in any food, that the virtues of olive oil can be as easily justified as the virtues of delicatessen products, the virtues of truffles as the virtues of potatoes. . . . Moral: Let us eat them all in balanced and moderate quantities.

Despite the mistakes sometimes made by the food industry, it remains true that our knowledge about food advances when nutrition is the concern of research institutions not dependent upon the food industry, and when the cook is enlightened by science. If the polyphenols in oil or wine do not have all the virtues attributed to them, they remain polyphenols nonetheless, sometimes capable of binding with proteins and contributing to the taste of dishes. If we recognize that the taste of cork comes not from one molecule— what is simple is always false, said Paul Valéry—but from many, then and only then does it become possible to ask how to eliminate it when using a corked wine in cooking.

EXPLORATIONS

Studies related to cooking cannot replace studies of cooking itself, which opens the way for science through the phenomena it presents, since science uses the experimental method to research mechanisms. Grilled meat browns: why? Poached lobster turns red: why? Heated egg white coagulates: why? Flour tossed into hot water forms lumps: why? For as many phenomena as exist, there are that many explorations, with as many possible discoveries at the end.

In those areas well illuminated by physical chemistry, new particles, such as the Higgs boson, will not be discovered, but the "cooking fool" will draw upon explanations proposed by science in order to perfect his recipes. From hors d'oeuvres to desserts, he will find the means to make a thousand technological transfers. Suffice it to say that there is something there for everyone. The scientist, saved by Knowledge from the bestial state that has us by the stomach, gleans from these studies new insights and new learning. The technologist, who prefers action and marvels at the

Panama canal (or today's massive bridges and space probes sent to Mars), will use the new information to innovate, to create new dishes and perfect old ones. The technician, who, not content to make others act, takes action himself, will head directly for the kitchen and will offer his dinner guests novel preparations or old dishes transformed in the light of new knowledge.

None of our dishes deserve to escape from these three approaches. The smallest phenomenon must be scrutinized by science, because the key to unsuspected gastronomic pleasures may be hiding there. Moreover, it must be said that scientific, technological, and technical criteria are "perpendicular." The new knowledge that astounds the chemist, physicist, or biologist leaves the technician or technologist cold if he does not yet see the practical possibilities. Conversely, a mechanism the scientist considers too standard or too simple can prove itself very useful to the technologist or technician, not to mention that technique sometimes goes hand in hand with art, and that Rembrandt needed only a piece of charcoal to make a work still worth stealing today.

AND TOMORROW?

In 1894, the chemist Marcellin Berthelot predicted—it is true that this speech was given at a banquet where there was much drinking—that in the year 2000 people would be eating entirely synthetic nutritive tablets. Furthermore, he predicted the end of war! His "tomorrow" is our today, and sadly, wars continue one after another. Luckily, the first prediction has proved as false as the second; we are not reduced to nutritive tablets, and better still, an analysis of cookbooks shows that not only has agronomy overcome famine (not universally, it is true), but it has also improved our foods.

Since I hear grumbling from those nostalgic for the food of their youth, I will hurry to make my point. Once again I am posing the question of "what will we eat tomorrow?" The best way to tell the future is to anticipate it through action: We will eat tomorrow what we learn to cook today! Already the advent of molecular cuisine—which was that historic moment of technological transfer accompanying the start of systemizing that chemical art which is cooking—has brought to our plates dishes concocted with

utensils not previously existing in kitchens, used upon ingredients never previously employed.

That is a beginning, but we can do much better, if only we will open wide
the doors of science. Tomorrow we will eat what our imaginations will allow
us to discover. Let us dream . . . yes, but let us dream effectively, supplied with
the knowledge necessary to give that dream substance. Let us dream actively,
equipped with a good provisions kit, before entering the real kitchen, an exercise that will prepare us for approaching the cuisine of tomorrow.

BEFORE SETTING OFF

Before leaving to discover a new land, we must have sufficient provisions
for the journey: money, food and drink (an essential here!), and so on. Exploring the physical chemical world of cooking raises the question of the knowledge of dishes, ingredients, transformations. Ideally we ought to have all of
chemistry and physics in our bundle for this voyage into the great culinary
world, but none of us are that rich.

And so we set off with our little bags anyway, audacious enough to think
we will find en route what we need to continue. The texts that follow thus
comprise "inns" or "guest tables" where places will be set, where the most
important phenomena will be presented (diffusion, osmosis, condensation, Maillard reactions), but it is still useful to do a bit of packing before
we leave.

Let us remember that, as a rule, matter is made up of "phases": gas, liquid, solid. The gasses are hardly eaten at all, amusing though it may be to
consume a food filled with helium and hear oneself talk like a duck for a
few moments (since the speed of sound is different in helium, the sound
spectrum produced by the vibrations of our vocal chords alters our voices).
Liquids? In cooking, there are two principal types, designated as "water" and
"oil." The cook knows that water is almost banned in cooking, because of its
serious fault: taste . . . or lack of taste. In the texts that follow, water is not
"pure" water, that dangerous liquid that makes absinthe cloudy, as Alphonse
Allais said. Oil? Despite those ads for the nutritional virtues of olive oil, it
would be very sad not to have it at our disposal.

So what are water and oil? The physical chemist designates as "water" any liquid composed primarily of water, any "aqueous solution." Thus, for the physical chemist who thinks (not the one who drinks!), wine is water, as are orange juice, bouillon, tea, coffee, and so on. Similarly, melted foie gras joins melted chocolate, melted cheese, melted butter (browned or not) in the category designated as "oil," where, naturally, we also find almond oil, walnut oil, olive oil, pistachio oil, hazelnut oil, and so on. As for the solids, they are very hard, to be consumed as a whole, but let us not forget that they remain solids when they are cut up or broken down into powders, strips (julienned), and so on.

In cooking, what are these phases made up of? One doctrine would have them be protides, lipids or glucides, sometimes mineral salts. Ridiculous. Foods are first of all made of water, actual water, with its molecules made up of hydrogen and oxygen. And while it is accurate to say that many food molecules fall into the three categories listed above, those terms are chemically obscure for those who know them only through hearsay, and insufficiently precise for those who know them intimately.

WORKERS OR ARTISANS

For instance, by protides, we understand amino acids as well as proteins. Let us look at an example. An egg white is made up of 90 percent water, as is demonstrated by the experiment that consists of gently heating it on a radiator or in a nonstick frying pan. Its initial mass, about 30 grams, is reduced to 3 grams, which appears as a solid, yellow sheet. This sheet is similar to packaged gelatin sheets in its look and its composition. In both cases, it is made up of protein molecules.

Protein molecules? Let us think simply of the strings on which pearls might be strung. According to the particular case, these strands are folding back on themselves (rolled into a ball) or not. The pearls? They are the amino acids just mentioned. But then again, no, the pearls must only be called amino acids when they are still in the form of independent pearls, and they must strictly be called "residues of amino acids" when they have reacted, because it is true that, having lost atoms during condensation, which binds them together, chemical entities strictly identical to the initial ones no longer exist. In our journey to the country of culinary dishes and cooking, however, we will not make those distinctions. Let us march straight on and not concern ourselves with the bumps in the road, so as not to lose time!

Oh, I forgot! There are two kinds of proteins. Organisms are composed of one kind; they are the bricks. Enzymes, the other kind, are the workers. Thus, the alpha amylase, an enzyme present in saliva, "hydrolyzes" the starch in flour to produce sugars.

THE FATS

The lipids? Here again, the category is not what it is often believed to be. Food industry advertisements often claim that fats are made up of fatty acids, and they beat us over the head with saturated, unsaturated, monounsaturated, polyunsaturated, omega 3, omega 6 . . . until they make us believe that these fatty acids are in fats.

Not true! Above all, oil contains triglycerides, molecules with a chemical structure that has been elucidated by the French chemist Michel-Eugène Chevreul. The structure? Let us think of a comb with three teeth; the spine of the comb would be glycerol, and the teeth fatty acids. More specifically, a glycerol residue and fatty acid residues, since, once again, the molecules have lost atoms to bind together.

Let us add that it is incorrect to liken lipids to triglycerides. First, the fatty acids (they also exist in a free state) are lipids, and second, the phospholipids are an important class of lipids. Indeed, they are a vital class, since all living cells are confined by membranes made of an assembly of these phospholipid molecules. Their structure? Let us imagine an electrically charged head with two "legs" that are two fatty acids. The head is "hydrophilic," which means that it situates itself easily in water. The two feet are "hydrophobic," which means that they must be given a lot of energy if they are going to mix with water molecules, and consequently, that they assemble instead with molecules of their own kind. That is why phospholipid molecules, like the lecithins in egg yolk, put into water, spontaneously form micelles, with the tails gathered at the center of these kinds of spheres that present the hydrophilic heads to the water, or why phospholipids in living organisms assemble in double layers, the legs facing toward the center of these double layers, the heads thrust into the water.

WITHIN TODAY'S SCOPE

Now for the glucides. This category is vast, even immense, and heterogeneous. The simplest sugars make up part of it; let us cite glucose, which is the fuel for our own cells (it is funny to think that we are collections of cells, all individually alive, but speaking together, is it not?) and which is also present in the majority of our foods. Or fructose, which is widely present in vegetables. Sucrose, or table sugar (why do we remain so fearfully attached to sucrose, even though we can use glucose or fructose, with their different flavors, just as easily in cooking?), is a slightly bigger molecule since it is composed of the first two. And, step by step, we can imagine larger and larger glucides, up to amylose and amylopectin, which constitute the starch granules in flour. In plants, these two molecules are assembled, glucose after glucose, into long chains, into many-branched trees.

The category of glucides is a very diversified world. In addition to amylose and amylopectin, let us not forget cellulose (if we want to see it in a pure state, we can look at the hydrophilic cotton in our medicine cabinets); the hemicelluloses; the pectins, which make jams jell; chitin, which is the polymer present in the shells of prawn and in mushrooms . . . and countless other "polysaccharides"!

Mineral salts? Trace elements? And all the other molecules? My old friend Jean Jacques, who spent his whole career as a chemist at the College de France, often complained that "in high doses, chemistry is such a pill." Let us not focus too closely on the details without due cause, for fear of offering the authoritative course on chemistry, a tiresome business, because it serves no purpose. If the case arises, the following pages will provide the necessary information. Without a second to lose, let us move on to cooking and science.

EXPLORATORY WORKS

Let us observe the marvelous phenomena that are revealed by culinary endeavors during those chemical or physical transformations brought into play by the cook. On this subject, I must report a personal error. In previous writings, I let myself go so far as to claim that cooking is chemistry and physics. I retract that statement, I confess my sin, I smudge my brow with ashes, I deplore the meanness of my spirit! Yes, because cooking is not chemistry, it is not physics. Cooking is a technique, a practice, that produces culinary dishes. And chemistry is a science, that produces — exclusively — knowledge. A culinary dish is not knowledge; it is not the mechanism science proposes to make sense of a phenomenon.

Moreover, I denounce a certain confusion between the science of chemistry and its applications, also often called chemistry. No, it is not chemistry that made the city of Toulouse explode a few years ago, and it is not chemistry that gassed the French soldiers in Verdun or Ypres in World War I. It was people who were responsible in both cases, and it is too easy — and cowardly — to lay the blame on a scientific discipline. Likewise, however much Marie and Pierre Curie explored the structure of the atom, they are not responsible for Hiroshima. If confusion reigns, we must remove it with the right words. I propose calling the applications of chemical science "molecular technology." And we must also give a specific name to the applications of physics, biology, and so on.

We are thus, once again, on safe ground. The facts are that cooking is a practice, which brings into play phenomena, which are studied by molecular gastronomy, which is a specific branch of physical chemistry, unless it is a specific branch of chemical physics; that hesitation on my part is proof, is it not, that Science is all one, without neat boundaries that can be easily drawn?

In short, transformations take place in cooking (we cut, we heat . . .), and phenomena are observed: The soufflé rises (when things go well!), the mayonnaise "takes," the béarnaise sauce "binds," the egg coagulates. . . . Each phenomenon merits an analysis, a scientific study. And for any number of phenomena, there are an equal number of studies. How to find our way in the myriad of studies? How to impose a little order? Let us remain in the kitchen or at the table, which are familiar places, with menus that distinguish the appetizers, the main courses, the side dishes.

By "appetizers," we mean not just the usual hors d'oeuvres, but also scientific ones, involving great principles that deserve to be known — and better, used — by all cooks. For example, if it is true that a puff pastry shell, sweet pastry, or strudel can be produced by rolling a ball of dough thin enough to make a transparent sheet, it is also true that understanding the power of the exponential function is essential for arriving at the laminated dough much more rapidly!

Once those principles are established, there are the phenomena particularly important to cooking: the browning of meats that we roast, the diffusion of odorant or sapid molecules, the changes in color. . . . Now we are at the "main course." This is the heart of cooking, which we must explore to really understand the extent to which cooking is a central cultural activity.

Finally, it is especially necessary to keep our eyes open, on the lookout for the smallest stimulating observation regarding the least important culinary preparation . . . to understand that culinary empiricism, if it has not been able to construct a coherent intellectual edifice—and how could it, given its empirical nature—has nonetheless brought to light a thousand phenomena along its erratic path, which, no doubt, would otherwise have gone unnoticed.

One example: pear compote, which, according to some cooks, turns red when cooked in a tin-plated copper pot. Experiments show that, no, pears do not turn red when cooked in such a pot. So why this "culinary detail"? Studies have shown that most pears can be made to turn red not by adding red wine when cooking them, but by understanding that their polyphenolic compounds (not to worry; the term is explained in the pages that follow) give pears color according to the acidity. What yesterday's cooks had been observing was that the more acidic pears (though, in the mouth, this acidity was masked by their sweetness) reddened, and they had—wrongly—attributed the change in color to the tin, which, it is true, makes red fruit turn purple!

Cooking is full of these strange historic twists and turns that provide science with grist for the mill. Naturally, there are "minor details," but it is up to us to find in them the seeds of general principles important to cooking . . . and why not to science? We will see especially that the question of diversity in sauces has led to establishing a system analogous to chemistry's, but for complex dispersed systems (formerly called "colloidal"). And we will see how such a system leads not only to scientific studies but also to many practical applications.

Yes, the engineer who "is always thinking" (as Louis Pasteur said) can make the most of any phenomenon, of all knowledge, in cooking as elsewhere—but here, in cooking.

WHAT WILL WE EAT TOMORROW?

We have seen how the fantasy of the nutritive tablet, the pill, haunts us, at least since that speech by French chemist Marcellin Berthelot. Berthelot did not understand the great lesson of evolution: Phenomena among living beings are dictated not by chemistry, but by the millions of years that have imposed their law. Existing as individuals, we are, above all, representatives of a "successful" species, successful, at least, in reproducing and surviving, despite predators, at the expense of prey.

Our sensory apparatus was primarily formed for that end: to escape predators, capture prey, find sexual partners with a view toward ensuring descendants. Bertholot's great mistake was to have neglected this point and, more importantly perhaps, to have wanted to place chemistry in the dominant position, whereas science must be "in the service of." In the service of humanity and its culture (what makes us human!), because it provides meaning, offers intelligibility; in the service of industry, in that the knowledge generated can be applied by technology.

All this is to say that, if making predictions is quite risky, there are facts. On the one hand, the pills and tablets are fantasies, fears. On the other, we saw the consequences of the mad cow crisis as professional cooks, who are attached to their aspics, their Bavarian creams, rejected gelatin, used for millennia, to adopt new jelling agents, which they had cursed in the 1980s! No one had predicted this change, though molecular gastronomy nevertheless desired it (today, the mad cow crisis seems to be over, and the cook's "piano" is thus enriched with new "notes").

Predict the future? Let us not tempt fate, but look instead to see how we could help alter the future of cooking. Of course, these are ecological times, and we could imagine that this great movement in civic consciousness might be opposed to the use of molecules in cooking. Nevertheless, economic law takes precedence, and if wine growers in certain regions are going to pour their excess wine into the river, we can also imagine that the fractionation of excess wines, to the extent that it produces polyphenols useful in cooking, will be appreciated by everyone, growers and cooks alike. Have you tasted these polyphenols? Do not hesitate, and report back to me, if you are truly a gourmand. They are so easy to use. You just add a few pinches. . . .

In the following pages, we will also see how mathematical calculations can lead to an infinite number of new dishes, which refutes the idea of Brillat-Savarin, according to which the discovery of new dishes contributes more to the happiness of mankind than the discovery of a star. No, if that were true, it would be too easy. Then we will see that the claim of an infinity of new dishes is no exaggeration; on the contrary, it is three infinities of new dishes, more precisely, to which molecular gastronomy has led!

The most important question is this: Which of them to produce? Practically speaking, which of these new dishes will cooks produce? And why those? At this point, it is essential to recognize that cooking is, first of all, a matter of love, since the goal is to make the dinner guests happy. That idea is an avowed hope, not a universal rule; for Tallyrand, cooking was a matter of power; for others, it is a matter of money. Nevertheless, if we retain the most generous idea, most conforming to the amiable spirit of gastronomy, then we have at our disposal a "filter" for choosing which new dishes will merit our attention . . . unless, since it is a matter of love, this filter is spelled with a "ph"!

There is also the question of art, not to be overlooked. In the following pages, we will see how cooking could be nonrepresentative, abstract. In other writings, I have proposed that it can be fauvist, cubist, impressionist, neo-Impressionist. . . . Yes, cooking can be all that, if we can only manage to detach ourselves from the "tradition."

Tradition, transmission: The tradition is what has been transmitted to us. Steak and fries, leg of lamb with green beans, hard-boiled eggs with mayonnaise, pike with béarnaise sauce, sauerkraut, beef stew, cassoulet, brioche, buckwheat cakes, pâté, terrine. . . . I like to compare all that to classical music, after which have followed so many beautiful compositions; Debussy is not Mozart, but far from diminishing our musical pleasures, he has, on the contrary, increased them. What if "culinary constructivism," which abandons the tradition, were a new way of increasing pleasures? What if we finally agreed to eat new dishes?

The consequences would be clear; the "tradition" that we transmit to future generations would be enriched by the work of our epoch. Is it not a small price to pay to increase both the knowledge of culinary phenomena and the repertoire of cooking?

I

Let Us Play with Our Senses

Sensorial physiology is a science concerned with perception. Sight, smell, hearing, touch, balance. . . . For each sense, there are stimuli and the organism's reactions, interpreted by that marvelous organ, the brain. This science seems very far removed from cooking because, with its necessary reductionism (breaking down phenomena in order to study them), it necessarily distances itself from the eater and, a fortiori, from the cook, who is concerned with culinary transformations.

Jean-Anthelme Brillat-Savarin, the author of *The Physiology of Taste*, which has traversed languages and centuries, wanted to establish a physiology of taste, to the point of claiming the role of "professor" in his book. To tell the truth, his book has only the appearance of science, the title and a few sections setting forth what a mind lacking science's experimental method can say about observable phenomena. Not that *The Physiology of Taste* is a bad book, quite the contrary, but it is a book on gastronomy, not physiology. Gastronomy is "the intelligent knowledge of whatever concerns man's nourishment," as Brillat-Savarin so accurately put it. Intelligent discourse, and not the production of dishes. Our great elder assembles a mass of heterogeneous facts, a priori knowledge, observations, anecdotes, since it is true that anecdote is the sauce that binds the pieces of meats, the indispensable spice for whetting the appetite to learn. . . .

In short, Brillat-Savarin, who was adviser to the French Supreme Court, after having served as his province's representative to the Third Estate, was a magistrate, not a scientist. Nevertheless, this book did much to help develop an interest in and knowledge of sensorial physiology in particular.

Taste is a synthetic sensation. Sight is important, as we will see, and hearing as well, although its influence is less well known. Smell has undoubtedly been overestimated, but science has made progress, and in recent years, studies in molecular biology have succeeded in identifying the olfactory receptors. The tastes? Due to a priori knowledge, the tastes raise some red flags . . . but times are changing.

The great Antoine-Laurent de Lavoisier explained that science cannot be perfected without perfecting language and vice versa. To make progress in the science of taste, new words will no doubt have to be introduced (*sapiction*, for example, to designate the perception of sapid molecules), and other words (*flavor*) killed off, in order to control the Tower of Babel problem in sensorial physiology!

In Praise of Superficiality

Our eating is no farther away than our tongues. And the superficial impression is not only the deepest one; it is often the only one.

The difference between a successful mayonnaise and a failed mayonnaise? It is a matter of how the very same molecules are arranged. In the case of the failed mayonnaise the water molecules supplied by the egg and the vinegar are at the bottom of the receptacle, with the oil on top; in the successful version, the oil molecules are distributed in droplets in the aqueous phase. This organization changes everything; in the successful case, our mouths come into contact with the water, but in the failed version, they come into unpleasant contact with the oil.

Other examples in which the structure determines the taste? Let us compare heated water into which butter is whisked and heated butter into which water is whisked. With equal parts butter and water, the results, however restored to the same ambient temperature, are very different. With the butter-in-water preparation, we obtain a creamy substance, because although the fatty droplets harden in cooling, they remain separated by a film of water; this is called a suspension. On the other hand, in the water-in-butter preparation, the consistency is firmer, because the butter forms a continuous phase in which the water is dispersed. At the ambient temperature, the proportion of the butter's triglyceride molecules in the solid state is about 5o percent, enough to constitute a solid architecture, held together by the stacking powers of the triglyceride crystals. Different cohesion, different consistency.

THE FUNCTIONS OF COOKING

Let us continue our analysis with a beef roast. What purpose does cooking it serve? Clearly, the heat kills pathogenic microorganisms that colonize on the meat's surface, but cooking seems something of a gustatory handicap because it toughens the meat, coagulating the proteins that are in the muscle fibers of which the meat is composed. This analysis is incomplete; roasted or grilled meat takes on flavor . . . on the surface, and only on the surface. Certain reactions known as the Maillard reactions, and many others as well (oxidation, hydrolysis, etc.) give the crust a strong roasted taste. In the mouth, we first perceive this taste and do not realize that the interior of the meat has the same taste as raw meat (which it is, when the meat is cooked "rare").

THE NOBLE ART OF CARVING

This observation explains the gastronomic difference between carving a leg of lamb French style and English style. The first is cut perpendicular to the bone; the second, parallel to it. The slices of meat offer different arrangements of muscle fiber; the first case is called *entrecôte*, with the fibers perpendicular, and the second, *onglet*, with the fibers parallel to the surface. The English carving method provides homogeneous slices, however rare or well done, whereas the French carving method provides each guest a bit of the brown surface, well done and tasty, and a bit of the rare interior, juicier, with more of the sui generis lamb taste.

SURFACE AND DEPTH

How to use the idea that the surface is important in cooking? Let us begin with a cube of meat, and let us cut it into three slices, the central slice being two times thicker than either the superior or inferior slice. First, let us cook the central slice in bouillon at a low temperature (about 60°C) for a very long time in order to dissolve the collagenic tissue that makes meat tough. After several hours of cooking, the meat should become as tender as butter. Next, let us heat the two other slices of meat in order to evaporate the water and toughen them. Then, let us put the three parts back in place; through this arrangement, we have thus obtained what roasting does less discretely, and we are certain of having a crisp crust on the surface and a tender interior. By way of comparison, let us test an arrangement, from bottom to top, of half a tender layer, two hard layers, and another half tender layer.

We can obtain the same effect with a sweet example by comparing two strips of chocolate with mousse in between them and two strips of chocolate, stacked, inserted into a mousse. In the first case the teeth hit the hard strips, and we feel their hardness despite the mousse; in the second case the teeth sink into the mousse until they reach the hard strips. Each time, the ingredients are the same, but they are arranged differently . . . and, in addition to the more or less appetizing appearance, we perceive first the taste of the superficial part, which determines our perception. "Only superficial people do not judge by appearances," warned the witty and insightful Oscar Wilde.

Those Delectable Fats

Fats are delicious because they are (also) associated with a sensation of heat in the mouth; the fattiest products seem the hottest.

Fats are a bane to the health of westerners, and all the more formidable for being adored. Why are we so fond of them? First of all, because they dissolve odorant molecules; in the refrigerator, poorly wrapped butter "takes on odors." Thus, fats trap scents.

The captured odors are restored when the fats are consumed; when foods are warmed in the mouth, the odorant molecules are released, because the bonds between the dissolved molecules and the fats serving as "solvents" are easily broken. Rising into the nose, the freed odorant molecules contribute powerfully to the taste of foods; hence our appreciation of fats. Additionally, fats lubricate the mucous membrane, contributing to a pleasing unctuous sensation.

FAT AND HEAT

At the Wageningen Centre for Food Sciences in the Netherlands, Jon Prinz and his colleagues are now exploring a third detection mechanism of fats: the sensation of heat that fat molecules induce in the mouth. Their work follows from a discovery made in 2003. Tasters given mayonnaise samples containing different levels of fat noted differences in temperature, although all the samples were at the ambient temperature. The fattiest mayonnaise samples seemed to be the hottest!

The Dutch physiologists repeated this experiment. With trained tasters, they tested crème anglaise and mayonnaise samples containing different proportions of fat (three levels, between 0 and 3 percent, for the crème anglaise, and with total oil content between 1 and 82 percent for the mayonnaise), stabilized at a single constant temperature. These studies have confirmed a very strong correlation between the perceived temperature and the percentage of fat.

To evaluate the precision of the physiological system for temperature detection by the mouth, the physiologists used small heating systems, with temperature sensors, and they discovered that the lower lip is the part of the mouth that is most sensitive to variations in temperature (0.08°C), variations that are felt in less than a tenth of a second! The lower lip is more sensitive than the front third of the tongue (sensitivity of 0.26°C) or the middle third (1.36°C). Pastry cooks are right to test their fondants and other products with the lower lip.

RECEPTORS AND EVAPORATION

The detection of hot and of cold is done by two different systems of receptors, so that the initial question becomes two questions: We must explain sensitivity to heat, on the one hand, and sensitivity to cold, on the other. The latter can be explained by the health advantages that led our ancestors to drink clear, fresh water from streams and rivers, rather than tepid, stagnant water from ponds.

In contrast, how did heat sensitivity arise in mammals? The detection of heat in the mouth undoubtedly has a specific function linked to food; some have proposed that this sensitivity might have facilitated the mammal's search for its mother, but wouldn't sight and smell have been sufficient? And why would the capacity to detect heat persist into adulthood?

Among many animals that do not sweat, such as dogs, the body's thermal receptors measure the humidity of the air and regulate evaporation by the oral mucous membrane, which cools the animal (in the same way, we are chilled by the cold when the wind sweeps over us as we emerge from a dip in the ocean).

The Dutch physiologists have proposed that foods containing more water seem colder than foods rich in fats, because the fats coat the mucous membrane and prevent evaporation and thus cooling. (The evaporation of volatile compounds, such as ethanol, would prompt an even greater cooling.) The detection of temperature by the mouth might be a useful mechanism for detecting fats necessary to the diet, and such a mechanism would be an advantage in adaptation.

This hypothesis has yet to be demonstrated, but in any case, as cooks, we can take into account the thermal effect of fats for dishes that we want to make lighter. We know that peppery and spicy products have the same receptors as heat. Through their action on these receptors, could such products compensate for the actual cooling caused by the evaporation of water in a food that lacks fat?

Flavor Unworthy of the Name

We are discovering the subtleties of olfaction, but we lack the words to describe gustatory perceptions. Using the word flavor is an error . . . in taste.

Gastronomy is also the art of speaking about the taste of dishes. Unfortunately, we lack the words, and the use of English is obscuring the French vocabulary. *Goût, saveur, arôme,* even *flaveur.* . . . What do these French terms signify? Researchers at INRA (L'Institut National de la Recherche Agronomique) in Jouy-en-Josas are providing the keys for understanding olfaction; their advances, as well as neurophysiology, show that the word *flaveur*, from the English *flavor*, has no place in French.

PEPPERY, SWEET, SALTY: YOUR SACRED BALMS

Taste is a sensation . . . a gustatory one; the sensation one experiences when one eats has many components. Let us bring some food toward the mouth. First, our eyes show us its form and color; visual sensations are an integral part of taste. The most recent proof is the experiment carried out by the Bordeaux Institute of Enology (see "An Enological Slant" below), in which tasters described the taste of a white wine, colored red, with the words used for red wines, because the sight of red gave them the taste of red wine in the mouth. (The added colorants did not change the taste of wine tested blind.)

Tactile sensations play a part as well, but our culture and the widespread use of packaging has made us forget that touch, apart from in the mouth, is a component of taste. If our fingers discover small crystals on fruit jellies made from beets and sugar, and then sprinkled with crystallized sugar, we will experience it as blueberry or black currant fruit jellies.

Then we bring the food toward the mouth and we perceive its odor, which results from evaporation of food molecules. The more volatile these odorant molecules are, the greater number of receptor cells they stimulate in the nose.

Here physiologists have made progress by discovering the intermediaries between odorant molecules and receptor cells in the nose: the OBP (for odorant-binding proteins, that is, proteins that bind odorant molecules). First identified in insects, these proteins that bind to odorant molecules before leading them to the olfactory receptors have been discovered in the human organism. In 2000, E. Lacazette and his team found human genes analogous to the insect genes that code these proteins. At the INRA Center in Jouy-en-Josas, Loïc Briand and his colleagues demonstrated that these proteins are expressed in the nasal mucous. In vitro, synthetic proteins, copied from a human OBP, bound to numerous odorant molecules from the class of aldehydes and fatty acids.

Let us return to the question of vocabulary. The nose, which captures odorant molecules, perceives odors. Those odorant molecules are sometimes called "aromas," confusing them with the sensation they engender; that is clearly a mistake, because a sensation is not a molecule. Moreover, preparations composed of odorant molecules judiciously blended are improperly called aromas.

Among the molecules evaporating in this way, certain ones do not stimulate the olfactory receptors but rather the receptors linked to a fascicle of nerve fibers with three branches, called the trigeminal nerve. That is the case with the menthol molecule, present in mint, which has an odor and which also communicates the sensation of coolness; to the olfactory sensation is added a so-called trigeminal sensation.

Now the food enters the mouth. Some of its molecules pass into the saliva and are then bound to molecules, called receptors, on the surface of special cells in the oral cavity. These molecules, called "sapid," give the sensation of "tastes." The cells that bear the receptors of sapid molecules are grouped together into papillae (the little rounded zones that we can see on the tongue). These are commonly called "taste buds" in English. But since we call the perception of taste "gustation," what should we call the perception of tastes? I propose "sapiction."

Warmed and broken down by mastication, the food also lets odorant molecules evaporate in the mouth; those molecules then rise along the back of the mouth toward the nose, through the retronasal fossae. This is still a matter of olfaction. In the mouth other food molecules act in different ways; some molecules stimulate the temperature sensors, others "excite." Cells or sensors detect mechanical characteristics; thus we perceive hard, soft, greasy, wet, and so on.

AN ORCHESTRA OF SENSATIONS

The whole of all these sensations, sapictive (tastes), olfactive (odors), physical, thermal, trigeminal . . . is the taste. Perceived by physiology, it is interpreted by the brain, which attaches qualities to it according to individual or social experiences (memories, emotions, training, etc.).

And what about flavor in all of this? Some food science specialists have proposed that the term brings together tastes and odor, but why combine these two sensations, since flavor is inaccessible? We cannot measure the sum of tastes and odor (at best, we can see it with the recent OBP findings, and we can begin to appreciate odors experimentally), and we can never perceive it, since our taste inextricably mixes tastes, odors, and other sensations. Flavor, neither perceptible nor measurable, is analogous to the angels that theology counts on the head of a pin.

Actually, the introduction of the word *flaveur* into French seems to result from a confusion with the English "flavor." Thus, according to the British Standards Institute: "Flavor: the combination of taste and smell . . . influenced by sensations of pain, heat, and cold, and by tactile sensations." Thus the French word *flaveur* seems to be a faulty translation of the English word "flavor," which translates into the French word *goût*—taste.

Let us stamp out that vile *flaveur*—we do have taste, after all!

That's Refreshing!

Physiologists have identified a receptor for molecules that, like menthol, give the impression of coolness. They will be able to synthesize "cooler" molecules.

Why are drinks with mint refreshing, even when they are served hot? At the University of San Francisco, D. McKemy, W. Neuhausser, and D. Julius have identified the DNA that codes a neuronic receptor upon which the menthol molecule acts, the molecule that is responsible for the refreshing effect of mint. Then these neurophysiologists identified the receptors activated by cold and they again found the one for menthol. An important discovery!

How do we detect hot or cold? When a hot or cold stimulus, a liquid that we are drinking, for example, passes a certain temperature threshold, specialized neurons in the mouth emit chemical signals that convey the sensorial information to the spinal cord and the brain. That is how the somatosensory system detects changes in skin or mucous membrane temperature.

A few years ago, neurophysiologists identified the vanilloid receptor VR1, present on the surface of tongue neurons and activated by capsaicin, the principal molecule in hot pepper, which we recognize by its "burning" taste. They then demonstrated that VR1 receptors were responsible for the perception of heat through the neurons at a moderate temperature threshold; then they observed that another receptor, VRL-1, that is analogous to the first but that does not react to capsaicin, is activated by temperatures above 52°C. These two receptors belong to a family of ducts called ionic ducts, through which ions—and thus information—pass in transit from the exterior to the interior of the neurons. D. McKemy and his colleagues then questioned whether other molecules of the same type participated in the perception of— not heat this time but—cold, with the idea that cold receptors were perhaps the same as those for the coolness of menthol.

This inquiry is more than academic. For decades, chemists have been modifying the menthol molecule to obtain molecules giving the sensation of coolness without the taste of mint. Unfortunately, these new molecules are difficult to synthesize and the sensory results not always convincing. Thus there was incentive for studying the receptors of the cooling action of menthol to better determine the nature of molecules that stimulate that receptor.

We have known for half a century that menthol and analogous molecules act upon the trigeminal nerve, the three branches of which irrigate the nose, mouth, and face. At the same time, physiologists have identified in mammals the small group of trigeminal fibers that discharge at between 15 and 30°C and transmit the sensation of cold to the brain, as well as the fibers that react at temperatures below 15°C. Other studies have shown that cold (about 20°C) prompts influxes of calcium ions in those neurons sensitive to cold. Thus physiologists suspected that the sensation of cold resulted from the opening of the calcium ducts. D. McKemy and his colleagues adopted the method used for discovering the vanilloid receptor and tested the reactions of neurons isolated from rats, to cold, menthol, and its analogues.

Having isolated neurons that react to menthol and to cold, the neurophysiologists recovered the segments of DNA used by the neurons for making the various proteins (the receptors are proteins) of these neurons. Then they introduced those DNA segments into cells derived from embryo kidneys. Finally, using a fluorescence microscope, they examined the modifications in the calcium ion currents in the cells after exposure to menthol. Our researchers thus identified the genetic sequence that codes the menthol receptor: a protein called CMR1 that belongs to the same family as the vanilloid receptor for spicy and for hot.

The question was posed once again: Do cooling molecules and cold act upon the same receptors? To elucidate this point, the specialists first used genetically modified cells to test reactions to eucalyptol, menthol, camphor, cyclohexanol, and icilin (about two hundred times more powerful than menthol), and they observed that menthol, eucaplyptol, and icilin act upon this type of receptor. Then they tested reactions to the cold in frog oocytes that had been transfected by the receptor. When they reduced the temperature from 35°C to about 5°C, a significant current of calcium ions was released. In other words, the same receptor is responsible for the reaction to mentholated coolness and to cold. Finally, D. McKemy and his colleagues have shown that the receptor is an ionic duct stimulator expressed in the small neurons of the trigeminal ganglia. Their original hypothesis was confirmed, all the more so when they observed that dogs tremble with cold when menthol is injected into their blood. The discovery of a strong chemical resemblance between the CMR1 receptor and a protein identified in the epithelium of the prostate explains this phenomenon. The article in *Nature* in which the three neurophysiologists present their work includes this note under the authors' names: "The three authors contributed equally to the published work." The scientific community understands from this note that the authors are aware of the importance of their discovery and that they anticipate the consequences of their work. Knowing the molecular structure of the receptor, chemists will be able to synthesize molecules capable of binding more specifically to that receptor than menthol or even icilin does, opening a considerable market, as the lovers of mint chewing gum know.

Tastes and Receptors

The discovery of a gustatory receptor for amino acids puts compounds that amplify tastes on the right track.

Molecular biology can identify the receptors of aromatic and sapid molecules; to a molecular question, a molecular response. Also, month after month, discoveries in the physiology of gustation follow one upon another. Cold receptors were discovered to be the same as those for menthol; C. Zuker, at the Howard Hughes Institute, and N. Ryba, at the Odontology Institute of Bethesda, then identified a protein that constitutes a gustatory receptor of amino acids.

Gourmandism is based on perceptions that are agreeable because they are vital to survival. The organism recognizes molecules it needs and toxic molecules that it must avoid. Sugar, which provides energy, is perceived as agreeable; alkaloids, often toxic, have an unpleasant bitterness (bitter tastes are only appreciated through cultivation; think of beer, which children do not like).

In addition to the tastes recognized by language, like sweet, salty, and sour, many others exist. For example, the taste umami has been recognized for a decade as the taste of kombu algae (kelp) broth, a taste owed to alanine and glutamic acid. Besides those two, other amino acids (the constituent molecules in proteins) are also reputed to have tastes, but their descriptions vary according to the individual.

It is not easy to determine the tastes of foods, because they cannot be separated from the odor, the appearance, the texture, the hotness, the cold. . . . To "approximate" the taste of amino acids, we can place on the tongue a mature, slightly salty cheese at room temperature, while using a small pump to blow a current of air into the nose (to eliminate the odor).

C. Zuker and N. Ryba identified the constituent proteins for the receptors of various sweet and bitter molecules in gustatory cells. They published in *Nature* their studies on the receptor named T_1R_{1+3}, which, among mammals, is activated by amino acids. Formed by two already known gustatory receptors (T_1R_1 and T_1R_3), the complex T_1R_{1+3} detects the L amino acids, those which we need to live, but not the D amino acids, symmetrical to the L amino acids in reverse, their mirror image, and not used by the organism.

Before their work, type T_1R and T_2R receptors had been identified. Gustatory cells equipped with T_2R-type receptors detect bitter molecules; those that have T_1R_1 and T_1R_3 proteins detect sweet molecules.

AMINO ACIDS REVEALED

Physiologists first studied the receptors in the T_1R family, which recognizes amino acids such as monoglutamate, gamma-aminobutyric acid (or GABA), and arginine. The receptors in the T₁R family are of many types and form various gustatory receptors. For example, we have seen that T₁R₂ and T₁R3 are sometimes expressed together (they thus form a receptor of sweet), as are T₁R₁ and T₁R3; only T₁R3 has been observed in isolation in gustatory cells.

Using embryo kidney cells, through gene insertion, physiologists have caused various receptors to be expressed in various combinations, beginning with cells that simultaneously express T₁R₂ and T₁R3. The reaction of receptors was detected by the movements of calcium ions (made evident using fluorescent markers) between the interior and the exterior of the cells. No L amino acid activated these genetically modified cells, whereas several D amino acids (which have a taste described as sweet by humans and are also pleasing to mice) unleashed notable cell activity. Then the researchers tested cells expressing either T₁R₁ receptors or T₁R3 receptors; they did not react to the L amino acids either. On the other hand, the simultaneous expression of two T₁R₁ and T₁R3 receptors prompted an intense reaction with the L amino acids, while the D amino acids provoked no reaction.

Having thus observed a detection of amino acids, the physiologists pursued the investigation by testing other molecules that engender a sensation analogous to umami, like inosine monophosphate (IMP). In rats this molecule provokes a gustatory activation detected by electrical registerings of gustatory nerves. In the experiments with renal cells possessing receptors T₁R₁+3, the reaction with L amino acids was considerably increased by IMP, which is thus a true "taste enhancer" (alone, IMP prompts no reaction). On the market, this molecule would meet with great success among gourmands!

Is the T₁R₁+3 receptor a umami receptor? Its reactions with many amino acids, as well as with sodium monoglutamate (MSG), seems to indicate that it is, but is it a principal or secondary receptor? The discovery leaves many questions open. For example, this receptor reacts with most L amino acids, even though all amino acids do not have the same taste. Some are equally pleasing to mice and humans, while others are neutral. Some are perceived as bitter by human tasters and rejected by the mice. Worse still, only a few amino acids have a taste analogous to umami.

With the help of cells that express the receptor, we will be able to study various molecules in order to understand what foods offer the umami-type taste. Sensorial tests have revealed that this taste is provided by cheese, meat, milk, tomatoes, asparagus, and certain seafood, but we will now be able to discover new molecules to replace "taste enhancers" like MSG or the inositides, now widely (excessively?) used by the food industry.

The Bitter Tooth

Teeth are indispensable to a good perception of the sweet, salty, and bitter tastes of foods. A single perception is lacking. . . .

Common sense is no guarantee of sound reasoning. On the pretext that we smell food twice when we eat (once when the food reaches the nose before entering the mouth; once when the odorant molecules released by mastication rise toward the nose through the retronasal fossae), it was decided that olfaction was the principal sense for taste. Moreover, the frequent loss of smell caused by colds made us believe that smell was everything and the perception of taste hardly anything.

Neurophysiological discoveries in the last decades are undermining this dogma. Why has it not been observed that the nonperception of tastes, exactly like the nonperception of odor, completely eliminates the sensation of taste? The experience of burning one's mouth on food that is too hot is sadly familiar . . . and demonstrative. And so? Sensorial neurophysiology shows that taste is a synthetic sensation composed of information of many different kinds: visual, olfactive, "sapictive" (the perception of tastes), trigeminal (hot, cool), mechanical. . . . And the loss of a single kind of information from the synthetic sensation is disastrous for taste recognition (which is the "goal" of the gustatory system that, shaped by evolution, aims to ensure the recognition of food).

"A single person lacking, and it is all unpeopled," lamented Lamartine. All the perceptions are necessary to be able to obtain the taste of a food. Does this idea hold for the consistency of a food as well? Annick Faurion at the Jouy-en-Josas INRA and his colleagues at the universities of Paris and Tours researched the influence, if any, of dental treatments on taste.

Dental treatments, especially the extraction of teeth, are frequent. In a prospective study, D. M. Shafer and his colleagues at the Odontology School in Phoenix analyzed taste deficits among half the patients who had had their third molars removed. These deficits resulted from damage to gustatory nerve fibers. Similarly, it was observed that dental surgery can diminish gustatory sensations for as long as six months, as a result of compression, stretching, or edema in the lingual nerve.

In order to learn if dental treatments cause a loss of gustatory sensitivity, A. Faurion and his colleagues considered 387 healthy, nonsmoking subjects, under no medical treatments; their mouths were x-rayed if necessary. The subjects were divided into groups according to the number of dental extractions or treatments of the dental canal, and their sensitivity to tastes was determined using an electrogustrometric technique. This consisted of stimulating little zones on the tongue with the help of a small electrode. Under the effect of the current, ions contained in the saliva were put in contact with papilla receptors, and the threshold of sensitivity was the smallest amount of current giving rise to the perception of taste. Nine zones of the tongue were tested: the tip; the anterior edges where the density of so-called fungiform papillae is maximal; the dorsal part, left and right, where the density of those papillae is minimal; the edges of the tongue; and the posterior part, both sides. For each determination of threshold, the electrode was placed on the tongue and, after a time of adaptation to the parasitic mechanical stimulation, a current was applied for one second. The subjects had only to say that they perceived a sensation. (The subjects did not know if a current was applied or not.)

The results are unequivocal: The greater number of unafferent teeth (or teeth without nerves), the higher the threshold of gustatory sensitivity (that is, sensitivity is diminished). It is not a question of age; no matter what their age, subjects having more than seven "dead" teeth have thresholds of perception higher than those of other subjects. Inversely, according to age group, no statistical difference was observed for subjects having fewer dead teeth. Moreover, an association was observed between the localization of deficits in taste perception and the position of extracted or treated teeth. The highest thresholds for anterior sites, unrelated to an anterior injury, showed that the neurophysiological convergence of somatosensory dental paths and sapictive paths could be responsible for the lessening of sapictive sensitivity.

Thus, the loss of taste in older individuals does not seem to result from a disappearance of taste buds, as is often believed, but rather from tooth loss and poor perception of the consistency of foods. To remain creatures of taste, let us keep our teeth!

An Enological Slant

Gourmets describe the bouquets of wines by evoking objects that have the same color, white or red, as the wines being tested.

Let us listen to the specialists who describe the bouquet of a wine in terms of fruits, flowers, plants, minerals. . . . In the thousands of years that they have been singing the praises of wine, they have not found qualifiers that are intrinsic to it; all the adjectives refer to products that have nothing to do with the grape! Let us stress this point: For other perceptions, we have specific words—sweet, bitter, salty, sour for taste; deep, shrill, muffled, among others, for sound; rough, smooth, soft, hard for touch. . . . For the aromas of wine, nothing.

Who does not know the story of the great enologist unable to distinguish a white wine from a red wine blindfolded? That unfortunate incident is true, and we have seen worse. Among twenty of the world's best bartenders, we have distributed nine glasses of white wine, nine glasses of red wine, and two glasses of ethanol in water; in the dark, neither sniffing nor noncomparative tasting let them identify what they were served! Gil Morrot of the INRA site in Montpellier, and Frédéric Brochet of the School of Enology at Bordeaux University have researched such phenomena.

To this end, Morrot and Brochet first analyzed the remarks of tasters. They based their work on five bodies of testimony: three thousand comments on tastings given by the enologist Pierre Dupont in *La lettre de Gault and Millau*, between 1991 and 1996; three thousand of the thirty-two thousand comments on tastings (selected at random) published by the *Guide Hachette du vin*; seven thousand comments in English offered by Robert Parker in the *Wine Advocate*; two thousand comments from a French wine grower.

Beginning with the complete text of these commentaries, a computer program, Alceste, did a statistical analysis of the occurrence of words, reduced the words to their roots, eliminated rare words, divided the texts into shorter units (the length of a sentence), and then divided the units created in this way into lexical groups.

One result is clear: The olfactory descriptors relating to red wines are generally associated with dark or red objects (chocolate, musk, tobacco, cinnamon, dark cherry, etc.), whereas the olfactory descriptors for white wines are principally drawn from light or yellow objects (plum, mango, honey, lemon, etc.).

The researchers then corroborated the lexical analysis using a study focusing on two bordeaux wines (A. O. C. Bordeaux, 1996): a white wine produced from sémillon and sauvignon vines, and a red wine obtained from cabernet sauvignon and merlot vines. A portion of the white wine was colored red using anthocyanin from grapes, and the gustatory neutrality of the coloration was first tested by fifty individuals; anthocyanin has no perceptible odor, and the colored white wine was indistinguishable from the uncolored white wine.

Then the three wines (white, white colored with red, red) were tested by fifty-four students of the School of Enology at Bordeaux University, under standard tasting conditions. The students were not informed of the experiment. Over the course of the first session, only the red wine and the white wine were presented, and the students described the wines using the terms of their choice. Then during the second session, each student received one white wine and one white wine colored with red, which they had to describe using the terms they had chosen during the preceding session; these terms were given to them in alphabetical order, and for each of the two wines in question, they had to indicate, by percentage, each descriptor's relevance.

THE TASTE OF WHITE?

The results of the two sessions confirmed the lexical analyses. The students first chose descriptors corresponding to dark or red objects for the red wine, and descriptors corresponding to light or yellow objects for the white wine. Then during the second session, the students generally used the olfactory descriptors they had chosen for the red wine in order to describe the white wine colored with red, and for this same wine they eliminated the descriptors associated with white wine; to describe the white wine, they retained the descriptors that they had previously chosen for the white wine.

The facts are obvious. The color of a wine determines its olfactory appreciation, and gourmets are under the spell of a sensory illusion. They taste the wine, perform a conscious act of sensory (olfactory) determination, and articulate a perception by using descriptors, but that perception is determined by the color and not by the olfactory perception, strictly speaking. The analysis of this illusion is leading to other tests that reveal other cases in which color influences the verbal description of the olfactory perception. For example, old wines that have taken on an orange tinge with age generally have orange, dried fruit, or chestnut aromas and rosé wines are often described by evoking pink fruits (currants, strawberries, etc.).

The reason for this illusion? In 1999, analyses of cerebral activity by means of positron emission tomography demonstrated that the cognitive processing centers for odor activate the V1 zone of the primary visual cortex (a zone that plays a part in the processing of visual images, the identification of objects, and the construction of mental images). Visual information, such as color, leads to the formation of an image in which this color is present. Moreover, one zone of the primary visual cortex, the left cuneus, is specifically activated during the task of verbally describing odor.

These analyses may perhaps explain why human beings have not coined specific terms to describe odors; if the identification of an odor derives from a visual process, it is logical that odors are identified by visual identifiers.

Scents, colors, and sounds respond to one another, wrote Baudelaire in *Les Fleurs du mal*. By evoking the correspondences between scents and colors, the poet was on the path of a neurological truth.

A Gustatory Illusion

The simultaneous contrast of impressions is transposable from colors to tastes.

Famous for his works on the composition of fats, the French chemist Michel Eugène Chevreul (1786–1889) is known in the art world for two reasons: First, the photographer Nadar devoted a photographed interview to him on the occasion of his centenary; second, he contributed to the development of the neo-Impressionist school of painting with the discovery of his "law of simultaneous contrast of colors," put into practice by Seurat, Pissarro, Delaunay, Signac. . . .

In cooking, this play of contrasts is equally possible. To convince you of the effect of contrast, look at the illustration opposite: The vertical gray band in the center appears darker than the two vertical bands on either side of it, although they are all the same shade and color.

PRICKLY PIGMENTS

Let us return to our hero Chevreul. At the end of a long study, Chevreul discovered that colors influence our perception of their neighboring colors: A blue spot on a white background seems to be bordered by yellow, and two colors like red and green seem to "vibrate" when they come close to one another.

The contamination of the neighborhood by a color is physiological. Yellow does not appear on the white paper because of yellow paint that might have colored it, but (and Chevreul did not know this) because the photoreceptors of the retina are influenced by neighboring photoreceptors.

In effect, we see thanks to three kinds of pigments, which capture different wavelengths. White results from the activation of pigments of the three types. If photoreceptors specific to blue receive wavelengths associated with blue, they are activated, sending the brain the signal that blue has been detected, and they inhibit neighboring neurons of the same type, in the zone that detects white. The receptors that capture blue, in the zone that receives white, are switched off, and only the other two types of neurons signal their activity to the brain. Thus the brain receives information according to which blue's complementary color (the white less the blue) has been detected. This is the "simultaneous contrast of colors." From which is derived the interest in complementary colors in painting and in color arts.

A SIMULTANEOUS CONTRAST OF TASTES?

Does such a phenomenon exist in cooking? Undoubtedly not, since taste is not the result of detection by three types of receptors, but rather a synthesis of many senses: sight, touch, smell, perception of tastes. . . . Might a weaker version of the law of simultaneous contrast of colors exist for taste? Let us return to our illusion of juxtaposition of one color, a blue of intermediate intensity, with dark blue and white (considered as a blue of very low intensity). The intermediate blue seems paler when it is near the saturated blue, and darker when it is near the white.

For a culinary transposition, we only need to similarly assemble three identical food masses (bland ones, for example) in which odorant molecules (1-octen-3-ol, with the odor of the forest floor) or sapid molecules (salt, sugar, etc.) are dissolved according to three different concentrations. In the first mass the concentration of odorant or sapid molecules will be nil; in the second mass, the concentration will be perceptible but weak; in the third mass, the concentration will be high. In taste tests the intermediate concentration will seem more concentrated if it is tasted after the bland or insipid mass, but less concentrated if it is tasted after the very concentrated mass.

In practice, we can do this test with fromage blanc and sugar, but after such a good start, why stop there? Chevreul actually collaborated with artists! Invited to put this effect into practice, the cook Pierre Gagnaire proposed a cream pastry dessert, made with milk, beaten egg yolks, sugar, and flour. He heated the mixture to the boiling point, dividing it into three masses to which he added bitter almond extract in three different concentrations. He let the three masses cool, then added whipped cream to them, and distributed the three cream desserts on a plate, sprinkling them with slivers of fresh almond. A fine way to appreciate the gustatory illusion!

2

Health and Diet

Health and diet. . . . The duo is a land mine! Because of the obesity pandemic that rages today in certain countries. Because of a certain industry that "invents" benefits for health. Because of the fear of eating unknown products. Because of new human behaviors, new industry and business organizations, new capacities for mixing molecules.

Must we eat ten fruits and vegetables a day? Tomorrow, the quota will drop to five. And the day after that, a vegetable or fruit will become less important than exercise. Bread? It is prohibited, but here it is back on the table again. Wine? It is responsible for France's alcoholism, but now it is good for the heart if consumed in moderate quantities. We are even seeing guinea hen and delicatessen products extolled for their alleged nutritional virtues!

Rereading my *Kitchen Mysteries*, I still stand by my conclusion: Priority must be given to the works of science, like those of many of my colleagues at INRA, that are establishing the facts. Retrospective epidemiological studies? They make me think of the story of the Martian whose flying saucer hovers over Orsay Station. The Martian sees a gentleman coming along the platform, then another one, then a third, then a whole crowd . . . and suddenly a train arrives; the crowd boards. The phenomenon is repeated, again and again. The Martian's foolish conclusion? That crowds make trains arrive!

No, the observation of a certain longevity or a reduced incidence of cardiovascular diseases (a huge category, indeed even a media "catchall") and increased consumption of one ingredient (strawberries, chicken, tomatoes, who knows what?) is not sufficient to establish that that ingredient is "good for the health." Moreover, if the solution is found in diversifying our diets, we should understand that no single food is "good for the health."

Let us be patient. Let us encourage the human nutrition research centers that do prospective epidemiological studies. Let us ask science for studies spanning decades. During that time, let us avoid the grossest eating errors, not forgetting that some facts are already established.

For example, satiation results for many from the stimulation of taste receptors. To satisfy our dinner guests and keep them from "stuffing" themselves, let us create meals that have taste . . . and allow the time necessary for the appetite to be satisfied (about 20 minutes). In practice, let us serve one soup (or another), with chunky bits in it, but let us construct the soup so that it stimulates all the receptors and does not rush by in a flash. The "wisdom of the nations," adopted by Brillat-Savarin, already was saying, "You eat too quickly!"

Vegetables for the Bones

Consuming fruits and vegetables limits the loss of calcium resulting from a diet too rich in proteins.

Why do older people frequently break their hips? Why do so many women suffer from osteoporosis? Because they are deficient in minerals and suffer bone loss due to a lack of calcium. How to combat mineral deficiencies and the weakening of bones? Through diet. As calcium is the principal component of bones, nutritionists have studied the direct calcium contributions of milk products, but until recently, no major role was attributed to fruits and vegetables, probably because of their more modest and less available calcium content.

At the Theix INRA Station, M.-N. Horcajada, V. Coxam, and C. Rémésy demonstrated the effect of fruits and vegetables on the acidity of the blood. Their alkalinizing effect complements that of fibers (cellulose, for example), useful for preventing various digestive cancers. As we will see, the consumption of fruits and vegetables shifts the chemical balances in an organism in a direction favorable to less calcium loss.

Meat proteins have a less favorable effect. Even though meat consumption has been decreasing steadily for a century (the drop due to the mad cow crisis was only a temporary flux in a steady pattern), the western diet remains too rich in animal proteins. Animal proteins acidify the blood through their sulfurated amino acids, cysteine and methionine; in the blood, a portion of these amino acids deteriorates and that reaction increases the concentration of acidifying sulfate ions.

The bones of the skeleton limit this acidification by releasing calcium ions (and, in lesser proportions, magnesium ions), but in doing so, the bones become weaker. That explains the danger that animal proteins present. How much calcium does the organism have at its disposal, thanks to the skeletal system? The organism has little leeway. If protein consumption is doubled (the recommended quantity is a gram a day, according to kilograms of body mass), calcium loss is 1.75 milligrams a day, or 365 grams in twenty years. On average, the skeleton contains about 800 grams of calcium (the quantity of calcium is less for women than for men).

Fortunately, the diet provides calcium that restores those reserves. Another sort of palliative would consist of limiting the acidification, that is to say, enriching the diet with products that are basic. Fruits and vegetables contain organic potassium salts (citrate or malate), and these organic salts are transformed into bicarbonates, which alkalinize the blood. Could this knowledge be put to use?

MORE POTASSIUM, LESS SODIUM

In 1999, V. Coxam and his colleagues studied the effects of potassium in the form of citrates or chlorides, and they compared those effects to those of sodium in the same two forms. Combined with proteins, sodium has a harmful effect on the bones because it stimulates the urinary elimination of calcium, and its presence is often associated with strong concentrations of chloride ions, which acidify the blood and increase mineral loss.

These effects were studied using rats that were fed for twenty-one days according to three different diets. In the first diet the potassium to sodium ratio was high (14), as in fruits and vegetables; in the second diet, the potassium to sodium ratio was low (2), as in meats; in the third diet, the potassium and sodium were in the form of citrate or chloride ions. Physiologists measured the concentration of calcium (lost) in the urine daily.

The results are convincing. Calcium loss is maximal for the diet with high concentrations of sodium in the form of chlorides, and it is minimal with abundant potassium in the form of citrate. The final diet's effects on urine acidity confirmed the theoretical ideas.

A second series of experiments tested the long-term effects of potassium citrate using, in one part, ovariectomized rats, in a state of estrogen deficiency like postmenopausal women, and, in the other part, rats with normal bone and hormonal metabolism. This second study, which lasted ninety days, confirmed the first results: The measures of mineral density indicate that potassium, in the form of citrate, reduces the loss of calcium in the bones; the effect is identical for both groups of animals.

In sum, the nature of the anion, citrate or chloride, has a more important influence over the loss of calcium than the nature of the ion, potassium or sodium; citrate is better and potassium more favorable. Vegetables contain little or no sodium; their effects on the concentration of calcium in the urine thus depend upon how rich they are in organic acids, potassium citrates or malates. With few calories, fruits and vegetables provide potassium that restores the potassium/sodium ratio without imposing too severe a sodium restriction.

The question arises: Since our bones are better maintained if we consume more fruits and vegetables, how should we cook them? Besides classic English cooking, which makes them lose their flavor, and steaming, which is limited to (often necessary) tenderizing, can we discover cooking methods that give vegetables powerful tastes? A diet is only accepted if the recommended foods taste good. The ball is in the cooks' court.

Olive Oil and Health

Compounds have been identified in olive oil that prevent the ill effects of oxidizing agents.

If we are not what we eat (*Man ist was man isst*, is the German expression), the fact remains that certain foods are healthier than others. In high amounts the smoked foods consumed by the populations of northern Europe promote cancers of the digestive system, whereas the diet of the Mediterranean basin,

rich in olive oil, fruits, vegetables, and fish, is associated with a lower incidence of cancers as well as fewer cardiovascular diseases.

These findings result from the most extensive epidemiological study ever
conducted on the relationship between diet and cancer. Since 1992, researchers from ten European countries have been collaborating on EPIC (European Prospective Investigation into Cancer and Nutrition), a study involving
more than half a million subjects. Routine blood samples are stored in liquid
nitrogen for later analysis; subjects' measurements are recorded, as well as
their state of health. In June 2000 in Lyon, the first results were released:
Daily consumption of 500 grams of fruits and vegetables corresponds to a
decrease by half of the incidences of cancers of the aerodigestive tracts, and
a notable decrease in cancers of the colon and rectum; tobacco and alcohol
have disastrous consequences for rates of cancers of the upper aerodigestive
system. Thus the risk for one such cancer for someone who smokes a pack a
day is eight times higher than for a nonsmoker. Consumption of more than
60 grams a day of ethanol (about 75 centiliters of wine) multiplies the risk of
such cancers ninefold.

The extraordinary blood bank that was developed is also being used
for more targeted studies. Thus, in Heidelberg, Helmut Bartsch and his
colleagues at the German cancer research center DKFZ attempted to cor-
relate the consumption of several oils and DNA alterations in white blood
cells. They assumed, first, that the autooxidation of lipids (turning rancid)
prompted the formation of reactive compounds damaging to cells, and second, that those reactive compounds bind to DNA, prompting mutations that,
in short, engender cancers.

The German researchers collected blood from women who were par-
ticipating in the EPIC study and who continued to eat as usual. The women
who had the least altered DNA were the ones who consumed the most fruits
and vegetables. Thus the level of altered DNA is a good in vivo test of the efficiency of antioxidants, such as those identified in olive oil.

Why this interest in olive oil? Because epidemiological studies report a
reduced incidence of cancers and coronary diseases among Mediterranean
populations. Since the discovery of this correlation, many factors of the Mediterranean diet have been proposed to explain its benefits: the abundance
of tannins in the local wines, the diversity of foods, the low consumption of
smoked products, the high consumption of fibers, the significant consumption

of fruits and vegetables, the consumption of olive oil. . . . Olive oil is a "good" food product on several accounts. Like other oils, the majority of its molecules, the triglycerides, are composed of one glycerol molecule to which three fatty acids are linked. Nevertheless, more than 70 percent of its fatty acid is oleic acid, a monounsaturated fatty acid (with one double bond between the carbon atoms of the fatty acid structure), which is less sensitive to autooxidation than polyunsaturated fatty acids, such as the linoleic acid in sunflower oil. The oxidation of polyunsaturated fatty acids, on the other hand, engenders reactive molecules that bind to the DNA in human cells—we know this from in vitro studies—and that may increase the risk of cancer.

Other properties of olive oil may explain its nutritional value. In Heidelberg, R. Owen analyzed various olive oils in search of antioxidant compounds and demonstrated that those protective properties also come from the presence of many phenols (molecules all containing a ring of six carbon atoms linked to at least one –OH hydroxyl group).

As olives grow, between October and January, the principal phenolic compounds are bound to sugars, but as the fruit matures, enzymes separate the sugar from the phenols, which are freed. Many phenols were identified in olive oil, hydroxytyrosol and two lignans among them. All have strong antioxidant properties, and all are more abundant in extra virgin olive oils than in refined oils.

The study of autooxidation of oils shows that hydroxytyrosol is the most active compound; hydrogen peroxide (present in oxygenated water), which considerably reduced the viability of (cultured) epithelial cells of the intestine, is completely inhibited by the addition of hydroxytyrosol. R. Owen and his colleagues demonstrated that eighteen extra virgin olive oils and five refined olive oils had this same protective effect, whereas seven sunflower oils did not have it. The phenolic compounds in olive oil are powerful antioxidants.

Lignans were also studied for their protective effects against cancers. They seem to act differently. Their structure, similar to that of estradiol, makes them antiestrogens (which block the proliferation of breast cells). Olive oil also contains strong concentrations of squalene, a compound that is transferred to the skin (the sebum contains 12 percent squalene) and that may act against skin cancers.

Gastronomy urges the use of olive oil. Biology confirms that good taste and health go hand in hand.

Digestibility

Studies using marked proteins show that it is preferable to consume "slow" proteins, which are better assimilated by the body.

It is claimed that garlic is indigestible when the sprout is not removed, that mayonnaise is "heavy," that certain wines are heady. Does an experimentally measurable graduation exist for digestibility, from easily digested to completely indigestible?

In recent decades, physiologists have learned to measure the speed of gastric emptying, showing, for example, the different speeds of absorption for fats when they are in a pure state and when they are dispersed in water in the form of an emulsion (mayonnaise!). Physiologists are also identifying foods that pass through the stomach without being digested and which, fermenting in the intestines, cause flatulence, among other things. Nevertheless such measurements remain generalized; since they are not at the level of transformational exchanges and molecule degradation, they do not reveal the mechanisms responsible for observed effects.

TRACKING PROTEINS

At the Human Nutrition Research Center in Clermont-Ferrand, Bernard Beaufrère and his colleagues have studied the digestibility of various milk proteins, thanks to proteins in which one of the constituent amino acids, leucine, was marked with a carbon isotope. These proteins were absorbed by volunteers and they were tracked in the organism over the course of their digestion.

The physiologists wanted to know how the human organism uses proteins: how they are broken down into their component amino acids during digestion, how these amino acids pass into the blood, and then how they are either degraded or assembled into human proteins. It was known that the principal regulators of this metabolism are hormones, such as insulin, as well as the amino acids themselves. So the question arose: How does the metabolism of amino acids depend upon the foods consumed?

In 1997, the Clermont-Ferrand researchers explored this question with the help of units of milk containing leucine marked with carbon 13. First, they discovered that whey proteins are more rapidly digested than proteins in the class of caseins, which form aggregates in milk called micelles. This slower digestion undoubtedly results from the fact that casein coagulates in the acid environment of the stomach and that its proteins are less easily degraded by gastric juices.

That is how an unsuspected parallel between proteins and sugars came to light. For the latter, we have known for a few decades that table sugar and other "rapid" sugars greatly augment the concentration of sugar in the blood, whereas complex sugars like those of starch (with its chains of thousands of glucose molecules bound chemically) slowly and steadily increase that concentration. The same is true for proteins: The experiments made clear a distinction between "slow" proteins, efficient from the point of view of the metabolism (they lead to better synthesis of human proteins when they are absorbed regularly) and "rapid" proteins.

Eager to verify the first results, B. Beaufrère and his colleagues compared the speed of assimilation for amino acids and the speed of protein synthesis after the ingestion of various meals with identical amino acid compositions but different digestibility. Thus, a dish with (marked) casein was compared to a mixture of free amino acids composed in such a way that there were as many of each kind of amino acid in the casein and in the free amino acid mix.

For the other part, the physiologists compared a single absorption of whey proteins with several successive absorptions of the same protein, the total quantity of absorbed proteins again being equal in the two cases. The goal of this experiment was to study the difference in digestibility for a single intake of food and for meals divided into many installments.

LET US FEAST SLOWLY

The studies were carried out on healthy volunteers. Twenty-two young people prepared themselves for the experiment by all consuming the same meals and doing the same exercises for four days. Through blood samples and analyses of expired air, doctors determined the quantities of marked leucine in the blood and the efficiency of protein assimilation. The physiologists thus discovered that the concentration of insulin in the blood increased moderately an hour after the absorption of the mixture of amino acids, but not after the consumption of casein.

Likewise, the concentration of insulin increased after a big meal of whey proteins but not after meals eaten in installments. The concentrations of leucine increased after all the meals, but the mixture of amino acids or the quick intakes of whey proteins induced a stronger reaction than the two other meals. The theory of slow and rapid proteins was confirmed.

What to conclude? That slowly digested proteins favor their good metabolic utilization, one that leads to a good synthesis of proteins by the organism of the consumer. The gastronome concludes further that cheese is better for the organism than whey, and undoubtedly that cooked egg, meat, or fish are also better than raw egg, meat, or fish.

Thus, let us cook, and cook well.

Healthy Pigments in Wine

The molecules that may protect against cardiovascular diseases are the ones that give red wine its enduring color.

At the University of Strasbourg, Raymond Brouillard is studying the molecules that color red wines, pigments that may be implicated in the "French paradox." The latter, discovered about ten years ago, corresponds to a relatively low mortality from cardiovascular diseases in populations that nevertheless consume fats . . . but also (moderately) red wine. R. Brouillard is proposing a new paradox: The same pigments associated with cardiovascular protection ought to disappear over the course of aging. Whereas red wine remains red for years, even for decades!

PLANT COLORS

The pigments that engender the red color of red wines are anthocyanins; these molecules are present in numerous colored plants: orchids, petunias, red cabbage, morning glories. . . . The anthocyanins of the European grape vine *Vitis vinifera* are essentially composed of a polyphenol bound to a sugar.

The pinot noir, the vine from which most great wines are made, is the variety of *Vitis vinifera* that has the simplest anthocyanins.

These polyphenols have a structure composed of three carbon rings, two of which are aromatics (with alternating single and double bonds between carbon atoms), the third of which replaces a carbon atom with an oxygen atom; to this structure are linked one or many hydroxyl groups (–OH). The sugar is from glucose, bound by an oxygen atom to the ring that includes the oxygen atom. These *Vitis vinifera* anthocyanins are distinguished from other kinds in the *Vitis* type by the hydroxyl group they have in position 5, on the first anthocyanin ring; moreover, they include only one sugar.

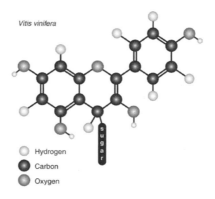

Vitis vinifera

○ Hydrogen
● Carbon
◐ Oxygen

A SECOND FRENCH PARADOX

The second French paradox stems from the enduring nature of the red color in wine. Studies have shown that in an aqueous medium, such as wine, anthocyanins are not stable over long periods of time because the water reacts with them.

For certain complex anthocyanins, the reactivity is weak (and the stability great) because the molecules are protected. Intra- or intermolecular copigmentations, which simultaneously alter the color due to anthocyanins, result, respectively, from the folding back of one anthocyanin upon itself or from the juxtaposition of two anthocyanins. The aromatic rings are piled up, which prevents the reaction with the water molecules of their environment. To these mechanisms is added a phenomenon discovered in morning glory blossoms: The color of anthocyanins is stabilized and altered when the anthocyanin is bound to cinnamic acid molecules.

The simple anthocyanins of *Vitis vinifera* do not seem to be protected in this way, because if intermolecular copigmentations are sometimes present, these copigmentations are less protective than intramolecular copigmentations and no stabilization by cinnamic acid takes place. If red wines remain red despite their reactive anthocyanins it is because their red is no longer that of the grape; the anthocyanins combine with wine molecules to engender new, more stable pigments.

Since the late 1990s, chemists have studied the existence of these combinations. As early as 1997, R. Brouillard and his colleagues showed that the first anthocyanin ring, the one that includes the hydroxyl group important to reactivity, is a key element in the formation of new pigments in red wine. Then, in 1999, the Strasbourg chemists used synthetic anthocyanins to identify the importance of this strategic hydroxyl group during the formation of another type of pigment with four hexagonal rings, which appears when red wines age. Next, it was discovered that pyruvic acid binds to anthocyanins and that acetaldehyde binds to catechin and malvidin 3-glucoside.

M. Rossetto and his colleagues at the University of Pisa demonstrated that the anthocyanins in pinot noir have molecular properties that seem to have a beneficial health effect: The protective properties have been tested in vitro and not in vivo, and the great debate is whether the cause for these correlations will be found. The benefit to health seems due to pigments specifically formed in red wine rather than to the original grape pigments or tannins. A good understanding of the chemistry of aging wines is leading to the elucidation of red wine's beneficial effects.

The Benefits of Royal Jelly

Fatty acids contained in royal jelly have unsuspected beneficial properties. Once again, advertising was right!

Nutrition and health. The theme is all the rage, even as the obesity pandemic spreads. A sedentary life style is an essential factor in the pandemic, but the promotion of physical exercise must also be backed by scientific research. Thus, products long renowned for health benefits are being reexamined in the search for useful properties. At the University of Athens, Eleni Melliou and Ioanna Chinou have studied the composition and properties of royal jelly. It is not a panacea, but many of its biological effects, due to fatty acids, have been confirmed.

Fatty acids? The reviews are full of them: saturated fatty acids, monounsaturated fatty acids, polyunsaturated fatty acids, and so on. What are they? Molecules composed of a hydrocarbonated "tail" formed by a chain of carbon atoms, bound to hydrogen atoms, and a "head" that is a –COOH carboxylic acid group. These various compounds play a variety of biological roles; they even serve as pheromones, molecules that ensure chemical communication in the animal kingdom, triggering mating behavior, for example.

On the other hand, fatty acids do not form oils and fats, as one might expect. These food products are composed of triglycerides, three fatty acids having reacted with glycerol. In this grouping the fatty acids and the glycerol have lost their individuality, exactly as the molecules in dioxygen O_2 and dihydrogen H_2 lose their individuality when they form water molecules H_2O.

The royal jelly studied by Greek chemists contains not triglycerides but isolated fatty acids. A creamy material, yellow, acid, with a light taste and subtle piquant odor, it is secreted by the mandibulary and hypopharyngeal glands of bees. It temporarily feeds the workers (for less than three days), but serves as the exclusive food of the queens, in the larval and adult state.

ORIGINAL FATTY ACIDS

Over several decades, chemists and pharmacists have gradually demonstrated that royal jelly has antitumor, antimicrobe, vasodilatory, antihypertensive, disinfectant, antihypercholesterolic, and anti-inflammatory properties. Its most interesting chemical characteristic is its composition in fatty acids; modifications of the simplest structure, made only from a hydrocarbonated tail, lead to diverse activities. The free fatty acids in royal jelly have only eight to twelve carbon atoms. The main one is trans-10-hydroxydec-2-enoic acid, synthesized ex novo in the mandibulary glands of bees and found in no other natural product. Moreover, about half of the dry material in royal jelly is made of proteins, such as the principal proteins of royal jelly and royalisin, which have immunoregulatory and antibacterial properties.

Despite numerous studies, the chemical composition of royal jelly remains poorly understood. The Greek chemists separated out its various components and analyzed the chemical nature of separated molecules. Several dozen original compounds were thus identified, four fatty acids among them, and many odorant molecules, such as acetovanillone.

AGAINST MICROORGANISMS

Scientists have studied the antimicrobial powers of these molecules on cultures of microorganisms, bacteria, and fungi. This time it was no longer the benefits of royal jelly as a whole that were tested, but the effect of isolated molecules, more useful to the pharmaceutical industry. In particular, the tests showed the valuable antimicrobial properties of certain fatty acids. For example, sebacic acid is very antifungal; 3-hydroxy-dodecanedioic acid is a powerful antibacterial. This acid is quite effective against staphylococci.

The value for diet? The food industry is generally interested when new molecules are discovered in foods because it can then advertise that some of its products are "natural" in origin or "just like nature." The culinary value? It is unlikely that royal jelly will become a standard food item, but if its taste is pleasing, why not use its various valuable compounds—now that they are known—in cooking?

This proposal was made in 2005 (see "Is It Time? You Say the Word!" in chap. 7); the culinary trend that uses compounds individually in this way is developing under the name of "note-by-note cuisine." It demands a good knowledge of the molecules used, which is not impossible today; the example of the mythical royal jelly shows us that—and offers us happy surprises.

Gels for Lent

Mad cow disease has scared consumers away from gelatins derived from mammals. They could use gelatins from fish . . . warm water ones!

Even though gelatin is not among the products declared a risk by the food safety agencies, even though gelatin results from extraction processes that remove the risks (treatment with bases or acids), even though gelatin producers have shown that their gelatins are not infectious, the public remains suspicious; it fears mammal gelatins and seeks other jelling molecules for its gels, jellies, aspics, Bavarian creams. At the ESPCI (École Supérieure de Physique et de Chimie Industrielles) in Paris, Madeleine Djabourov, Christine Joly-Duhamel, and Dominique Hellio have studied fish gelatins and have explored the relationship between the jelling temperature of these gelatins and the water temperature where the fish live.

No doubt aspics date back as far as the cooking of meat in water. Heated to above 55°C, the collagenic tissue that is present around muscle fibers, in cartilage, and in skin is broken down. Initially in the form of triple helixes, the collagen molecules that make up this tissue are separated and dispersed in the water, taking the name *gelatin*. Then, when solutions containing gelatin molecules cool, the molecules reassociate along the extremities, forming segments of triple helixes. The strands cluster thanks to weak bonds, called hydrogen bonds, between a hydrogen atom of one gelatin molecule and an oxygen atom of another molecule, and a three-dimensional network is formed, trapping water molecules and dissolved compounds within the mesh of the network. The structure formed from water being trapped in a solid is what physical chemistry calls a gel.

In practice the preparation is simple: The broth in which meat cooks for a long time gels when it cools. Which raises the question that must be asked in times of food scares: If cooks are afraid of gelatin, why are they not afraid of the broth that provides that gelatin? Analysis shows that it is unknown gelatins, produced by the food industry, that prompt fear, despite studies that have shown that gelatins derived from contaminated cows are not infectious.

FISH, I TELL YOU!

Whether the fear is justified or not, the facts remain: The public is abandoning gelatins from mammals but would be ready to use gelatins from fish. So now the question: What gels do these gelatins form? Do they jell in the same fashion as mammal gelatins?

Let us eliminate one question: Yes, fish gels have the same absence of taste as meat gels if the extraction is done well. With her colleagues, M. Djabourov explored these questions within the framework of a European research program on the development of by-products. Fish gelatins were expected to be different from those of mammals because fish are not warm-blooded. Living in more varied environments, they ought to present more biodiversity. In 1956, for example, the Swedish chemist K. Gustavson observed that the denaturation temperature for collagens (the temperature at which collagens dissociate and the resulting isolated molecules form) increased with the temperature of the waters where the fish lived.

Then, in 1971, the American chemists W. Harrington and N. Rao explored the relationship between the jelling of collagens and variations in the amino acid composition of those proteins, essentially proline and hydroxyproline (a collagen molecule is composed of about one thousand amino acids). They discovered that the quantity of hydroxyproline in the sequence differs according to the collagens.

Altogether, these studies showed that it was no doubt necessary to compare gelatin extracts from cold-water fish, warm-water fish, freshwater and saltwater fish. The skins of these different fish were thus washed; then collagen was extracted using acetic acid. After purification using a centrifuge, the collagen was precipitated with salt.

Thanks to the studies done for many decades on bovine gelatins, the ESPCI research team knew that gelatin gels reach equilibrium very slowly; gelatin molecules take different times to develop into a network, varying according to the gap between the ambient temperature and the jelling temperature, but always a long time (more than 4 hours). These developments were traced using polarimetry; the rotation of the polarized light increases with the formation of triple helix segments. The physical chemists determined the jelling temperatures for various fish gelatins. Whereas beef and pork gelatins jell at a temperature of 36°C, tuna gelatin jells at just

29°C, sole gelatin at 28°C, and cod gelatin at 15°C. Thus the hunch that the jelling temperature would depend upon the temperature of the environment in which the fish lived was confirmed. The cod in particular is very much a cold-water fish! Additionally, the physical chemists showed that the formation of triple helixes begins at different temperatures according to the collagens, in an order that corresponds to that of the temperatures of fusion.

These results, which are valuable to the food industry, obviously have significant culinary consequences; a gel that forms at a lower temperature, let us remember, is a gel that holds up less well to reheating when cooked. So cooks, thus informed, will be well advised to use gelatin extracts from warm-water fish or to continue using those good old gelatins from cows or pigs.

Quality of Meats

The maturation of meat and its tenderness can be determined by measuring its electrical properties.

How can we be certain that the meat we are preparing to cook will be good? Cooks have their methods. For example, they examine the "marbling." When they see veins of fat running through the piece, they assume that the meat will be tasty because of the odorant molecules present in the fat, and that the meat will melt in the mouth thanks to fat's contribution to the meat's texture. To see if the meat is tender, they will also pinch it between the fingers: If the texture is close to that of butter, the meat has a good chance of being tender. Other cooks slide a thumb across it in order to feel the "grain." (Rough meat would be of doubtful quality.)

These are subjective, artisan methods that we would love to master. At the INRA Center in Clermont-Ferrand-Theix, Jacques Lepetit, Sylvie Clerjon, and Jean-Louis Damez have shown that the electrical properties of meat provide objective information on its state of maturation, an essential element in the mechanical resistance of meat, known as tenderness.

At the heart of the work is the idea of electrical impedance. For meat as for any other material, electrical impedance describes the way in which an object, placed between two electrodes, permits the passage of an alternating electrical current. This impedance is composed of two parts: On the one hand, electrical resistance, which measures the dissipation of energy in the form of heat; on the other, the capacity, which measures the quantity of stored electrical energy. In meat the capacitive component is significant because meat behaves like a condenser. Intra- and extracellular liquids, which contain ions, are conductors, but the movements of these electrical charges are hampered by the insulating biological membranes.

The INRA researchers took advantage of the variations in impedance according to the direction of application of the current (parallel or perpendicular to the muscle fibers). The cause of the variations is clear: Meat is an anisotropic material because it is composed of muscle fibers, which are elongated cells containing a network of proteins, water, and all that make these cells live. Those muscle fibers are restricted by their membrane and sheathed by a fibrous tissue containing collagen (from which comes gelatin, after long cooking in water). They are gathered into bundles by other tissues made of collagen, and fats are included in the structure.

TENDERNESS AND ELECTRICITY

How does tenderness develop in beef? The animal's state of health is important, but maturation after the animal is slaughtered is crucial. It is a slow process, and its duration varies considerably according to the animal. In France beef is generally put on the market after being stored one to two weeks, even though the optimum maturation is not always achieved; consumers pay dearly for meat that is not always tender. How do we satisfy them while minimizing the cost of storage?

Physical or biochemical laboratory methods already permit us to recognize this state of maturation, but these methods are not practical for the food industry. Researchers have thus perfected rapid, nondestructive methods of measurement, adapted to industrial constraints. They have discovered a relationship between the electrical and mechanical properties of meat over the course of maturation. During maturation electrical impedance of a muscle decreases proportionately to the mechanical resistance. Why? Because the muscle evolves over the course of maturation. On the one hand, the cellular membranes gradually deteriorate, which reduces their electrical capacity; on the other, the extracellular space evolves, which changes its electrical resistance.

Nevertheless the relationship they discovered between electrical impedance and mechanical resistance cannot be put to use in assessing the state of maturation, because the relationship between electrical impedance and mechanical resistance varies from muscle to muscle. A technological dead end? Not entirely. Very recently Theix researchers have shown, as we predicted, that electrical anisotropy (that is, the difference between electrical impedance measured parallel and perpendicular to the muscle fibers) is directly linked to mechanical resistance, independent of the muscle or the animal. This relationship results from a single biochemical mechanism that degrades the membranes and the network of proteins in the muscle fibers.

And as good luck multiplies, the researchers were able to advance another step with the coming of hyperfrequencies: When an electromagnetic wave with a total frequency between 300 megahertz and 20 gigahertz, polarized linearly (the direction of the electrical field is fixed) is transmitted, the dielectric anisotropy of the meat can also be measured. As with total frequencies between 1 and 10 kilohertz, the anisotropy of the reactions to hyperfrequencies diminishes over the course of maturation. Furthermore, with hyperfrequencies, it is no longer necessary to plant electrodes into the meat to register its impedance (which risks propagating microorganisms); the maturation of the meat is measured without contact, with the help of antennae.

Will hyperfrequency sensors soon be used to reveal the maturation of meat? When that time comes, butchers will know, better than by touch, how to determine the tenderness of their products.

3

What Are the Notes?

The "product"! Maurice-Edmond Sailland, also known as Curnonsky, a journalist who called himself the "crown prince of gastronomes," wrote that "things are good when they have the taste of what they are." Appalling assertion, aimed at authoritatively imposing a personal vision upon the collective field that constitutes cooking! No, things are not good when they have the taste of what they are, and moreover, the intellect cannot accept such a declaration without an explanation of the second half of it.

The fact remains that this idea has done much harm in the culinary world, which hides behind it either to perpetuate an old aesthetic idea or to justify the quest for "products." Because it is true that a tomato grown with enough light and nutrients has a more powerful taste than the anemic version of that same vegetable. It is true that asparagus withering in the market stalls is often worth less than asparagus fresh cut. It is true that button mushrooms just gathered are much superior to those blackened things left sitting around too long. And finally it is true that skate, when it is not fresh enough, has an unbearable stench of nitrogen!

Nevertheless, what are we really cooking? That is the issue here, more than the examples currently under hot debate. We will look at just a few examples, because the true question for cooking is how to make good use of the products.

The War of the Shallot

"Electronic tongues" can distinguish the shallot from the onion and pretentious hybrid varieties, which are thus not true shallots.

A brouhaha in agriculture: Producers of shallots, of true "traditional" shallots, are rising up against hybrid varieties of onions and shallots grown to be passed off as shallots. First of all, the facts: The Alliaceous family includes the genus *Allium*, for which the cultivated species are principally garlic, chives, scallions, onions, and leeks. If leeks, chives, and garlic are easy to recognize, confusion threatens to reign in the onion and shallot group, and even more so as plant breeders create hybrids! For the shallot, two types can be distinguished, pink shallots and gray shallots.

Now shallots, prized for the delicacy of their taste, are more difficult to grow than onions: Onions multiply through seeds while shallots multiply through bulbs. Bulbs must be removed from the shallot clusters and then planted by hand to form new clusters, from which bulbs will be extracted again for the following generation. It is three times more work to grow shallots than to grow onions, which accounts for their higher cost. Nevertheless, according to cooks, their taste justifies this additional cost. Could shallots be obtained from seed?

A SHALLOT IS A SHALLOT

Government regulatory agencies have failed to recognize the special nature of the shallot, which must propagate by bulb, even though the traditional shallot market has already been threatened by American surtaxes. The shallot producers decided to put up a fight, and as is often the case in these political economic wars, science was called upon to produce the facts. In particular, Claire Doré, at the Versailles INRA, and Gérard Sparfel, at INRA in Plougoulm, used an electronic system to learn if the shallot did or did not possess the specific characteristics attributed to it.

This study followed an earlier genetic analysis carried out at the University of Gatersleben. The German taxonomists Nikolai Friesen and Manfred Klass studied the genomes of onions, pink and gray shallots, and hybrids.

They first extracted genetic material from bulb cells, then they amplified isolated fragments beginning with DNA extracts, to search for differences between these fragments.

At the conclusion of this genetic analysis work, five groups clearly appeared: *Allium oschaninii*, *Allium cepa*, *Allium vavilovii*, *Allium asarense*, and *Allium pskemense*. All the gray shallots were grouped together with the reliably listed samples like the *Allium oschaninii*. The common shallot *Allium cepa*, an aggregate group, really does belong to the *Allium cepa* species.

SENSATIONS MEASURED

The use of an electronic tongue by the INRA specialists and their colleagues A. Legin, A. Rudnitskaya, and B. Seleznev in Saint Petersburg confirmed these results. Electronic tongues, similar to the electronic noses that were perfected a few years earlier, are devices consisting of a group of detectors (organic molecules bound to silicon chips) and an electric circuit that statistically analyzes the signals from the receptors when they are put in contact with a mixture of complex molecules.

Unlike electronic noses, in which the receptors are bound especially to hydrophobic molecules (odorant molecules are generally not soluble in water, which explains why they are released into the air), electronic tongues react especially to molecules in aqueous solutions, like the many sulfurated molecules present in the plants of the *Allium* genus. (That is why it is best to clean onions under a trickle of water, in which the sulfurated molecules are dissolved.)

When a sample is deposited on the network of receptors, they produce an electrical current that is analyzed by the electronic system. Corresponding to each sample is a point in a space that has as many dimensions (of current intensity) as the tongue has receptors. A statistical analysis program then examines how the points in this space are gathered together.

The tests done on onions, shallots, and various hybrids showed the following differences: The onion samples formed very distinct groups; the pink shallots made up one separate group, and the gray shallots constituted a third group, distinct from the first two; no onion was grouped together with the shallots and vice versa. Not only did the electronic tongue corroborate the genetic analysis, but it provided proof the regulatory agencies can use: It is enough to guarantee loyalty to and trade allowances for the shallot.

There was still the question of recognition among the actual concerned parties, the tasters. During an INRA seminar on molecular gastronomy, a three-way test was organized, opposing traditional shallots and hybrids. In the dark, the seventy some participants received numbered plates with two identical samples and a third different one; they were to say which two seemed to be identical samples. For products cooked or raw, the results indicated only a slight propensity toward good recognition. On the other hand, when appealing simply to sight and taste, the products differ.

The Scent of Things

The quality of strawberries is measured by networks of neurons. The taste of products can be distinguished according to the year, the soil, and the variety, as well as the significant effects of freezing.

How can we assess the quality of fruit objectively? At the University Institute of Technology in Périgueux, Michel Montury and his colleagues are using the results of their chemical analyses to objectively determine the taste of strawberries. Results: They can even identify the field where the strawberries were grown!

The taste of food products is determined by their odor and their taste (especially). To study odor, chemists use chromatography to separate the vapors emitted by the strawberry and they identify the components of the vapors with mass spectrometry. Odorant molecules are separated out in this way. All that then remains is to compare the peaks of various samples according to variety, dates and places of pickings, and so on.

A MIND FOR ANALYSIS

This comparison is difficult; because of the twenty-three molecules important to the odor of strawberries, and two or three samples for each of the seventeen strawberry varieties submitted, over sixteen hundred peaks have to be compared. Also, the chemists teamed up with specialists in "neuron networks," those mathematical tools that sort data without a priori knowledge. A good example of neuron network use is the recognition of numbers in addresses on letters: The network groups together the same handwritten numbers, distinguishing a badly written 4 from a badly written 7, according to criteria unknown to postal users but absolutely effective. The neuron network assesses that the variability of handwriting for a 4 is inferior to the difference between a handwritten 4 and handwritten 7, and separates them.

There are various types of neuron networks, but the Périgueux team used the "auto-organizing maps" invented by the Finnish physicist Teuvo Kohonen. Used for classifying data, these networks are composed of a layer of entry neurons that code the data and a layer of exit neurons that post the results. The entries here are vectors for twenty-three components (the intensities of the twenty-three compounds used for describing the odor of strawberries). The algorithm employed in these auto-organizing maps projects these data toward the exit neurons, and the data vectors that share the same characteristics appear close together on the exit map.

During the first study of strawberries picked in 1998 and frozen, the exit maps produced were divided into twenty-four hexagonal cells. All the samples of the same variety were found in the same cell, and the seventeen varieties were projected onto different cells. This discrimination demonstrated the value of the method: Each unknown sample was classified with the others of the same variety and there were no "intruders."

Thus equipped with an effective analytical tool, the chemists examined the annual variability of strawberries. This time, six varieties (Ciflorette, Cigaline, Ciloe, Cireine, Pajaro, and CF116) grown in 2000, 2001, and 2002, under the same conditions, were sampled and analyzed. A clear separation appeared for the 2001 and 2002 samples, demonstrating the importance of the growing year; the exit map also indicated the separation of varieties, but the interannual variability of a variety was greater than the intervariety variability.

STRAWBERRY SOILS?

Do strawberry soils exist, just as there are soils for wine? Six varieties grown all in the same year, in three different places, all in the Aquitaine, are well separated on the exit map. At a higher resolution, the three growing sites are separated for each variety; the variability according to production site is weaker than the variability according to variety, so that, for taste, the variety is more important than the place of production.

Calibrated in this way, the method of analysis allows gustatory questions to be studied: Is freezing harmful to the taste of strawberries? This time, eleven varieties were compared, in their fresh and frozen state. A rough map was sufficient to show a clear distinction between frozen samples on one side and fresh samples on the other. At greater magnification, the eleven varieties were distinguished in each half of the overall map.

In sum, selectors now know that the variability within a single sample is inferior to the variability according to the samples (the samples are very representative); the variability according to the samples is inferior to the variability of the varieties of strawberry; the variability of the varieties is inferior to the variability of the growing year. . . . The method that objectively assesses the difference in the taste of things can make us dizzy thinking about all its possible applications.

How Tannins "Melt"

As tannic wines age, they become less astringent because their tannins react chemically.

Throughout the history of discoveries, instrumental innovations enrich the purse of knowledge, and nothing gets done without them; a greatly improved boat was needed to discover America, a rocket to explore the moon and reveal its mysteries, spectrometry to analyze the molecules of complex mixtures. . . . Nevertheless the mechanisms of aging wine have remained obscure because of the complexity of the molecules that give it color and taste. For some years, Michel Moutounet, Véronique Cheynier, and their colleagues in the combined Sciences for Oenology department at the Montpellier INRA have used an improved kind of spectronomy to analyze phenomena linked to the aging and maturation of wines.

At the heart of their concerns are the polyphenols, molecules that play a major role in the taste and color of tea, coffee, olive oil, roses, and cosmetics,

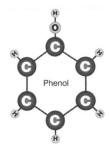

Phenol

as well as in protecting plants against insects. Many plant materials contain these molecules that include at least one "phenol" group, with six carbon atoms each bound to a hydrogen atom, except for one carbon atom that is bound to a hydroxyl group (composed of an oxygen atom bound to a hydrogen atom).

Tannins, molecules long used for their capacity to bind to the proteins of animal skins and so strengthen them, are polyphenols belonging to two families: hydrolyzable tannins, in which a sugar is bound to small polyphenols, and condensed tannins, which are large molecules formed through association of many units of a polyphenol called flavanol. These condensed tannins are also called proanthocyanidines because they release anthocyanins, the red or blue pigments in plants, when they are heated in an acid environment. In wines the polyphenols contribute to color and bitterness. Tannins are astringent; binding to the lubricating proteins of the saliva, they leave the mouth dry.

THE REACTIONS OF TANNINS

Why do red wines that age take on an orange tinge? And why do tannic wines lose their bitterness and astringency? It was thought that the polyphenols combined, forming larger and larger molecules, which lost their astringency and their bitterness, but later studies illuminated the structure of grape tannins and elucidated a few of the polyphenol reactions in wine.

The Montpellier chemists identified the structure of condensed tannins. In wines many polyphenols that do not belong to the tannin family are grape extracts. In addition, the hydrolyzable tannins come from the oak wood of the barrels or added enological tannins. And finally, condensed tannins are extracts from fruit skins and, to a small extent, from seeds.

Polyphenols are reactive molecules; their reactions are responsible for the changes in the color and astringency of red wines. Two principal types of reactions, enzymatic oxidations and reactions involving anthocyanins and flavanols (the subunits of condensed tannins), take place simultaneously during winemaking. The first type, similar to the reactions that make apples turn brown when cut and left out in the air, takes place especially at the beginning; it leads to products dark in color. The second type continues even when the enzymatic activity has diminished. These are the best-known reactions today.

They are important for the color because, in wine, which is acid, the anthocyanins are in two forms in equilibrium. One form is charged positively and red, and one form is predominantly colorless and hydrated; that is, the molecule is bound to a water molecule.

Many reactions involving the two forms have been proposed to explain the conversion of tannins into more stable pigments that give aged wines their orange tinge. Recent analyses have shed light on these reactions. As was assumed, the subunits of condensed tannins react directly with the anthocyanins. Moreover, the two types of molecules can react thanks to acetaldehyde, a small molecule produced by yeasts and by the oxidation of ethanol.

Finally, anthocyanins can be transformed into more stable pigments by reacting with the metabolites of the yeasts. All these reactions increase the size of the molecules, but the products formed are often unstable at the acid pH level of wines. Spontaneous breaks and reactions of the sort released with smaller phenolic compounds reduce the average length of the molecule chain derived from tannins. Thus the reactions in the wine on the tannins lead first to heavier compounds that split up into lighter compounds, with the prevalence of each dependent upon the initial polyphenol composition and upon conditions like the concentration of oxygen, the concentration of yeast metabolites (acetaldehyde, for example), the pH, and so on.

In sum, the analysis of phenolic compounds in red wine showed a reduction of the polyphenols rather than a growth of tannins; this reduction leads to the loss of astringency and the disappearance of bitterness. It is a happy coincidence and not an intuition that has led gourmets and enologists to speak of tannins "melting" over the course of the aging of wines.

The Power of Tannins Revealed

The tannins in wine are astringent because they bind to the lubricating proteins of the saliva. Interesting food/wine combinations can be drawn from measuring the strength of these bonds.

Let us (always!) begin with an experiment. Let us take a mouthful of astringent wine (it is called "tannic") and rinse it around in the mouth; then let us spit it out again into a clean glass. Scientific interest being stronger than disgust, let us examine the liquid. We will see in it precipitates resulting from the tannins bonding with the proteins in the saliva. That is why the mouth is "dry" or "puckered" after consuming such wines; precipitated in this way, the saliva proteins no longer do their task of lubricating.

Our observation thus leads to drinking tannic wines with dishes that contain proteins, in order that the latter, precipitated first, leave the mouth in good shape for tasting. To help us with these culinary combinations, H. Rawel, K. Meidtner, and J. Kroll, of the University of Potsdam, have measured the strength of bonds between proteins and phenolic compounds (the chemical class to which tannins belong).

MOLECULES IN VOGUE

These compounds are all the scientific rage. In recent years the *Journal of Food Science* has devoted more than one-third of its articles to them. There are reasons for this craze, both scientific (chemists are really beginning to analyze these compounds) and nutritional. (These molecules are antioxidants, which are often considered "good for one's health"; moreover, some are antimutagens or anticarcinogens.)

Nevertheless, these molecules sometimes have antinutritional effects resulting from the precipitation of proteins, which are then no longer assimilable. That is why, in a period of food shortages, the Corsicans who were reduced to consuming acorns (containing nutritious starch) added clay, which chelated the formidable astringent tannins in the acorns and avoided the precipitation of the proteins necessary to the maintenance of the organism. For the same reason, monkeys eat soil when, for lack of fruit, they eat leaves that contain tannins and other phenolic compounds in addition to the needed proteins.

The Potsdam chemists used phenolic compounds and proteins common in food: quercetin (present in numerous fruits, such as pears) and its glucosides (quercetin bound to glucose, which forms rutin and isoquercetin); gallic and ferulic acids (present in plums and cereals, and found intact in the blood after consumption of foods that contain them); and chlorogenic acid (in apples and other plant tissues) for the one part; gelatin (supplied by meat), caseins (from dairy products), alpha amylase (from saliva), lysozyme (in eggs, for example) for the other part; other proteins (bovine serum albumin) were chosen for their presence in the blood, through which pass the phenolic compounds assumed to be active in vivo.

How do phenolic compounds become bound to proteins? If it is easy to identify the possible covalent chemical bonds, it is more difficult to detect the weaker bonds; it is as if a fisherman had to catch fish with a hook from which the fish could detach themselves. So the German chemists used various methods, their fish being either proteins or phenolic compounds. First, they used chromatography to separate molecules from a mixture by making the mixture migrate into a gel; as the speed of migration depends upon the electrical charge and the size of the molecule, an isolated protein can be distinguished from the same protein bound to a phenolic compound. Other optical measures revealed alterations in the structure of the proteins (the way they fold in on themselves is disrupted by their association with phenolic compounds) and the association of phenolic compounds with other molecules.

THE BONDS OF SENSATION

The fishing was good. First of all, the bonds between proteins and phenolic compounds are either covalent for strong chemical bonds or noncovalent for so-called weak chemical bonds. In the case of weak chemical bonds, the methods used measure the strength of the bonds for various phenolic-protein compounds. For example, ferulic acid binds poorly to gelatin but binds well to human or bovine serum albumin, as well as to lysozyme.

Second, nutritionists will be interested to learn that albumin from blood serum binds well to the phenolic compounds studied; thus it is a kind of metabolic sponge for the serum. Third, alterations in environment such as the ones produced during culinary transformations (acidity, temperature, etc.) change the associations by modifying the proteins. We can begin to understand how cooking alters astringency, bitterness, and pungency in mixtures that pass through the saucepan.

Are we seeing the dawn of a true "molecular" cuisine, in which cooks will choose the taste of foods by mixing proteins and phenolic compounds they select in this way?

The Taste of Cork

A new method of rapid analysis for corks will allow us to detect the flaws in them that give wine a bad taste.

Defective corks are a plague; they disgust the gourmet, harm the innocent restaurant owner, wrong the wine producer whose products they damage, and disgrace the often helpless cork maker. How can we avoid this evil? At the AgroParisTech, L. Eveleigh and N. Boudaoud have perfected a method of rapid analysis for corks.

The taste of cork is an odor of mildew or wet cork tissue, more or less intense, that unfortunately does not diminish when the wine is aerated. Chlorine is often the cause of this taste; it reacts with the abundant phenols in wine and forms compounds called trichloranisoles, in particular. These molecules have formidable olfactory power: A tiny dose of trichloranisoles (5 billionths of a gram per liter) is enough to make the wine smell like cork. And the average amount in contaminated corks is as high as 5 percent!

STORAGE OR CORK?

The corks made from compressed cork seem more subject to "cork rejection" than others because the degradation of the glue used sometimes prompts the formation of trichloranisoles. Nevertheless, corks are not solely responsible for their taste; storage vats can also harbor molecules that give wine a cork taste. And finally, to complicate the scenario, trichloranisoles are not the only molecules responsible for the cork taste. Some chlorophenols have been suspected and found guilty, as well as many other organic molecules when their concentrations rise above a tolerable threshold. An example is para-ethylphenol, which gives an odor of old leather to aged burgundy wines when its concentration is lower than four parts per million, a peat taste to whiskies when its concentration is slightly higher, and a horrible taste of burned rubber when its concentration is higher still.

Today's analysis techniques allow the types of chloranisoles found in corked wines to be differentiated. When it is a matter of 2,4,6-trichloranisole, we can accuse the cork maker, because this molecule is typical of cork tissue, but when the laboratory identifies 2,3,4,6-tetrachloranisole, the environment is at fault, because this molecule is common in wood and the wine grower will have to inspect his vats.

To test entire lots of corks or corked wines, either the corks are soaked in water or the defective wine is diluted; then the volatile molecules are captured from the water or the wine with the help of a "trap" (a fiber in which these molecules will be absorbed); then chromatography is done on the gaseous phase, combined with mass spectrometry, after having dissociated the molecules derived from the chromatography.

These analyses have identified more than one hundred specific volatile compounds for the corks: organic acids like acetic acid, furans, aldehydes; phenols like vanillin; linear or branching hydrocarbons. Unfortunately, gaseous phase chromatography combined with mass spectrometry is a tedious process. So the chemists looked for more rapid analyses, using mass spectrometry of unseparated volatile products. Would statistical analyses of the data from this spectrometry be enough to classify the corks?

CLASSIFIED MOLECULES

The AgroParisTech chemists studied the question of whether statistical analyses of data from spectrometry would be enough to classify the corks. Tests were carried out on corks with three different geographic origins: Spain, Portugal, and Morocco. The first statistical analysis used was the analysis in principal components, which considers without bias the whole of the collected data. This analysis showed that the data associated with the corks was divided into three separate classes, corresponding to the three geographical origins.

A second statistical method, called least partial square analysis, also permitted these classifications to be anticipated. It was put to the test for new data, obtained through mass spectrometry analysis for the same corks after three and six months of storage. Finally the use of the two statistical methods simultaneously allowed particular fragments of the mass spectrometry analysis to be identified, in reduced number (three-quarters), upon which an identification of the corks' origins could be based. The chemists are currently trying to discover if these same methods could rapidly identify defective corks.

4

The Question of Hors d'oeuvres

Let us imagine that we must quickly set about cooking, without a book. What principles could guide us? Since culinary practice is the implementation of chemical or physical natural phenomena, let us seek our principles in those two disciplines.

In physics the question of the diffusion of molecules is certainly central; the molecules of gases and liquids are moved about haphazardly by bumping into their neighbors, transmitting heat, altering concentrations. . . . As a result, new possibilities for reactions arise. What chemistry is thus born? It is useful to distinguish the strong forces, which bind atoms in the "usual" molecules (for example, the oxygen and hydrogen atoms in a water molecule), and the weak forces, which bind molecules without causing them to lose their individuality, forming supermolecules.

The movement of molecules triggers chemical reactions. In cooking, the reaction identified in 1912 by the Nance chemist Louis-Camille Maillard is important: Sugars can react with amino acids to engender brown, sapid, odorant compounds. Nevertheless I am paying for my sins, because my appeal for the rehabilitation of Maillard has fostered the belief that Maillard reactions alone are responsible for the good taste of roast meat, bread crust, chocolate, and coffee. Not true! Chemistry is rich with a thousand other marvelous reactions that contribute to the taste of foods during culinary transformations. In a bouillon the hydrolysis of collagenic tissue engenders amino acids with specific tastes. Oxidation as well is important in cooking, beginning with the auto-oxidation of fats (or their turning rancid).

At this stage, I am aware that we have lapsed into chemistry, while I wanted to speak with you about physics. If diffusion is important, the notion of energy is central. Do molecules migrate? They are minimizing their energy. A chemical reaction takes place? Again it is a question of energy. Energy, I tell you!

Yes, energy, but also gels, to a lesser extent. A gel is a "dispersed" system, composed of an aqueous phase dispersed in a continuous solid phase. With such a definition, we understand that there are many types of gels. For example, in aspics, the gelatin forms a three-dimensional network that holds the water. Jams have the same structure; in them, the pectin holds the water. The white of a fried egg has the same structure. The same structure occurs in meats, fruits, vegetables, fish, composed principally of water "held" in the cells: muscle fibers for fish and meat; cells surrounded by a rigid wall made of cellulose, hemicelluloses, and pectins in fruits and vegetables. The difference is that these last gels are not "connected"; there is little communication between the cells, even though we know that the French specialty, the clafoutis, becomes stained the color of cherries after a short time, as the coloring molecules of the cherries migrate into the clafoutis gel through diffusion.

We are going in circles: diffusion, gel, diffusion . . .

It is time to begin on an exploration of hors d'oeuvres.

The Egg at Sixty-five Degrees

By cooking an egg in an oven at a temperature halfway between that of coagulation for the yolk and for the white, you will obtain a new egg.

An egg that cooks is one of those everyday miracles we no longer recognize; the transformation of a yellowish, transparent liquid into a white, opaque solid is a remarkable phenomenon, is it not? This is a thermal jelling. The proteins that constitute 10 percent of the egg white bind, forming a continuous network that traps the water of the egg white in a chemical gel. It is this jelling phenomenon that I propose we examine first of all.

The theory of how a gel sets has a prestigious lineage; so it was that in 1861, the Scottish physicist Thomas Graham proposed a classification of physical systems divided into aerosols, emulsions, suspensions. . . . Among the colloids (from the Greek *kolla*, glue), Graham included the gels that form hydrated silicic acid, hydrated alumina, starch, gelatin, egg white, and so on. At that time, jelling seemed similar to the crystallization of a substance beginning with a supersaturated solution, and there was no distinction made between natural substances, such as pectin or gelatin, and concentrated solutions of insoluble inorganic compounds, like barium sulfate.

Gradually physicists discovered that the state of gel was associated with the formation of a continuous network in the liquid. In the 1940s the theory of jelling progressed in two ways. First, the American physicist P. Hermans proposed a classification for the different types of gels (distinguishing the aggregates of spherical particles, the networks of fibers or of elongated particles, physical polymeric gels, and chemical gels made of supple threads linked by covalent bonds). Second, J. D. Ferry studied the composition of protein gels, in other words, coagulated egg white. Specifically, Ferry assumed that coagulation resulted from a double reaction: First, the proteins, balls wound around themselves, unraveled ("denaturation"); then the unwound proteins combined into a network ("aggregation").

The speed of these two stages determines the characteristics of the gel. Ferry proposed that if the aggregation is slower than the denaturation, the gels formed are less opaque and more delicate than the gels formed with a fast aggregation speed. In the 1970s in Göteborg, Sweden, Anne-Marie Hermansson tested these predictions by exploring the conditions that favor denaturation, such as an elevated or low pH; the electrical charges that the proteins thus possess favor interactions between the latter and the molecules of the solvent (that is to say, denaturation), but reduced aggregation. She confirmed that a more orderly gel is formed if the aggregation is slower than the denaturation, giving the denatured proteins time to orient themselves before aggregation; this gel is less opaque and more elastic than those for which the aggregation is not slowed down. Conversely, when the aggregation and denaturation are simultaneous, a more opaque, less elastic gel is formed.

IN COOKING, KEEP IT SIMPLE!

How can we use these theories in cooking? The cook who has mastered the preceding points risks being stymied by the complexity of the egg white, which contains 10 percent protein: ovotransferrin, ovomucoid, lysozyme, ov-albumin, and globulins. As for the egg yolk, it contains proteins bound to cholesterol (LDL and HDL), livetins, phosvitins. . . . What are the denatur-ation temperatures for all these proteins? Here again, the answer is unwieldy: the proteins are denatured at 61°, 70°, 75°, 84.5°, 92.5°, 70°, 72°, 70°, 80°, 62°, and above 140°C.

How can we get out of this predicament? Through experimentation. Let us put an egg white in a glass receptacle that is heated from the bottom. With the help of a probe we can measure the temperature at which the egg white, a yellowish, transparent liquid, turns opaque and hardens: about 62°C. The preceding data show that it is probably ovotransferrin that ensures this initial coagulation. For the yolk a temperature of 68°C will be obtained in the same way. At higher temperatures, when many proteins have coagulated, the consistency becomes harder because the networks associated with each coagulated protein hold the liquid phase better.

In an oven preheated to 65°C let us place a glass with an egg white, a glass with an egg yolk, a glass with the yolk and white mixed, and a whole egg in its shell. Let us wait a few hours (1 or 2 hours longer will not change the results, especially if the glasses have been covered with plastic wrap, which will prevent the water from evaporating and a crust from forming on the preparations); then let us remove the samples from the oven and observe them.

The white is set (since the temperature of 65°C is higher than the tem-perature of 62°C measured beforehand), but it is still milky, very delicate, and not rubbery as in overcooked hard-boiled eggs. As for the yolk, it is liquid. If the gamma livetin has a coagulation temperature of 61°C, its concentration is not sufficient to make the liquid set. And the whole egg in its shell is then peeled and poured into a bowl: a superb milky mass, coagulated but tender, with a perfectly regular form, with the yolk retaining a strong taste of fresh yolk and not the taste of omelet or hard-boiled eggs.

Finally, the glass containing the mixture of yolk and white is set, and we obtain perfect scrambled eggs, without lumps. The Paris chef Pierre Gagnaire has made a dish out of them, which he calls "oeufs brouillés de la Cité" (city scrambled eggs): Put egg yolk and white into a cup with a little sugar and va-nilla; place in a 65°C oven, and, when the mass is set, remove and serve with a slightly tart apricot sauce. Bon appétit!

Tahitian Fish

To preserve fish proteins, they are put into a solution, then jelled. In this process the bases "coct" better than the acids.

What is cooking? According to the dictionary, cooking is a transformation of foods that results from heating. Not a very precise definition. Is it enough to defrost a fish for it to be cooked? Certainly not, even though the food has been heated and transformed. And water that evaporates, is it cooked? *Omnia definitio periculosa.* Let us explore instead the realm of the cooked in order to define this notion. Cases of pathological cooking are eggs placed in ethanol (in cooking, plum brandy is preferable), which coagulate and form like poached eggs; or, again, eggs placed in bases (the hundred-year-old eggs of Asian populations) or in acids ("anti-hundred-year-old eggs," since bases are the opposite of acids). More common, Tahitian fish, soaked in lemon juice, an acid, is related to the South American seviche and surimis, versions of which, flavored with crab, have invaded much of Europe. At the Suranaree University of Technology, J. Yongsawatdigul and J. Park have explored updated versions of these preparations, obtained through jelling the proteins of the muscle tissue of a fish caught along the Pacific coasts: the yellowtail rockfish (*Sebastes flavidus*).

Surimis are traditional products that the wives of Japanese sailors created to extend the length of time fresh fish proteins would keep. (*Surimi* means "fish flesh.") After cutting off the head and gutting the fish, they ground the flesh into a paste to which they added egg whites, flour, oil, salt, and flavoring ingredients (spices, seasonings, etc.); the preparation, steamed, was sold in sticks.

PRECIPITATIONS AND COAGULATIONS

Modern production is a little different. The flesh is washed in large amounts of water, which concentrates the proteins that are responsible for the muscle fibers contracting; then it is coagulated (or jelled). Nevertheless the yield from this process barely attains 30 percent and, in 1999, an American team patented a process that makes soluble the myofibrillar proteins (responsible for contraction) and sarcoplasmic proteins (which allow all cells to live) and then precipitates them.

This patent is based on the fact that proteins are "spiked" chains of lateral groups, some of which are electrically charged. For example, the carboxylic acid groups –COOH of aspartic acid are negatively charged (they are ionized into –COO⁻) in a basic environment, whereas they are neutralized in an acid environment. The amino groups $–NH_2$ of the amino acid called lysine are positively charged in a basic environment (they are ionized into $–NH_3^+$). In sum, according to whether the environment is acid or basic, the proteins are charged positively or negatively. At the "isoelectric" point the total electrical charge of the protein is nil.

These ionizations determine the solubility of the proteins. In very acid or very basic solutions the proteins, charged in several places along the chain, repel one another and pass into solution, all the more so because the electrical charges have an affinity for the water molecules. At the isoelectric point, neutralization permits regrouping; the proteins precipitate, forming gels, in which the water is trapped.

TWO POPULATIONS OF PROTEINS

Chemists at the University of Oregon have calculated that myosin (one of the proteins responsible for muscle contraction) in salmon is soluble in very acid or very basic environments. Thus surimi was produced through neutralization of acid solutions of proteins, as had been done for many years, but also using basic solutions. (The neutralization is necessary for the product to be edible!)

Thailand chemists have shown that gels thus formed are firmer when they result from alkaline solutions, specifically because the enzymes known as cathepsins break down the myofibrillar proteins, even though these same enzymes do not seem to be active in an alkaline environment. On the other hand, the gels formed beginning with only myofibrillar proteins are firmer than those that also contain sarcoplasmic proteins.

Is this precipitation a kind of "cooking"? An email survey of 6,357 people registered on the "molecular gastronomy" list settled the question: 90 percent preferred the introduction of the word "coction," from the same Indo-

European root *kok* to the word "cooking" or *cuisson*, to describe this "cooking without heat." ("Decoction" is thus the later extraction of a food that has been "cocted.") So henceforth the results of the Thailand chemists will be stated in this way: surimis are coctable through alkalinization of acid solutions, but also through acidification of alkaline solutions, and the gels formed in the second case are preferable.

Take One Egg . . .

Let us reinvent the culinary repertoire. Let us discover an infinite number of new preparations using one simple method.

Is the cuisine we have inherited complete? Has empiricism enjoyed enough time to procure for us the totality of the dishes that our gourmand appetites have been waiting for? The question we are posing has no bearing on tastes, which are infinite, but on the possibilities for culinary transformations to produce "food structures" that will receive the tastes perfectly suited to them.

Let us begin with an egg, keeping in mind that we could do the same with meat, a vegetable, fish, or fruit. The Strasbourg philosopher Abraham Moles, inspired by Dmitry Ivanovich Mendeleyev's periodic classification of the elements, proposed "matrixes of invention." Let us follow his precepts. In the top part of a chart let us thus put the egg. Then, below, let us arrange squares in which to put the possibilities for division: the whole egg in its shell, or the yolk and white separated, but outside the shell, or again, the white and yolk beaten together, or the yolk alone, and finally the white alone (without forgetting the shell, which has been too neglected up until now).

DIVISIONS AND ADDITIONS

After the first operation of dividing the egg, let us decide upon adding another food component or not. Each square is divided into nine: one for adding nothing, one for adding a gas, an aqueous solution, an oil (a fatty substance in a liquid state), a solid, ethanol, acid, base (bicarbonate, or lye, which contains potash), and one, finally, for the addition of heat.

We have fifty-four squares, and some of them already correspond to new food products. It is clearly impossible to describe them all. Let us just begin. The whole egg, in its shell, with nothing added, is a raw egg that one swallows whole. The whole egg, in its shell, to which we add a gas, water, oil, is subject to no additional transformation. The same egg in a basic environment, like the ash, becomes a hundred-year-old egg, because in a basic environment the proteins are denatured. The same type of action takes place with acids, which leads to anti-hundred-year-old eggs. With alcohol? I invite you to do the experiment without delay. It takes a few months to obtain the results, which I call a Baumé, after the French chemist.

With the white alone, the addition of heat leads to results that depend upon the temperature attained, varying between soft-boiled and rubbery egg white; the addition of gas leads to the classic stiff peaks, while the addition of an aqueous solution leads to a thinned egg white, which we will "put to the side," as cooks say. The addition of oil produces a white emulsion, with the taste of the added oil. The addition of alcohol? The white coagulates, taking on a specific consistency. The addition of acid? It takes several days . . . but, no surprise, we again discover the coagulated white of an anti-hundred-year-old egg. With the yolk, the yolk and white separated, or together, we get the same kinds of results.

INVENTIONS GALORE

Let us go down a row, continuing to transform the results obtained. Thus the stiffly beaten egg white in the third row could be heated; have added to it a gas, water, oil, solid, base, acid, alcohol; or be heated once again. What if we beat the egg white, adding a liquid to it? The white takes on volume, and if we do it right, we can obtain as much as a cubic meter of foam that will have the taste of the added liquid: lemon if we have added lemon juice, coffee if we have added coffee, bouillon if. . . . And if we cook the beaten egg white? And if we add oil by whisking it in? And if. . . . I leave the surprises to you, and propose that you descend one row farther down the chart.

Each square in the fourth row again creates nine new squares for the fifth row, which leads us to a total of $6 \times 9 \times 9 \times 9$, or 4,374 squares. Not all of them are relevant, but many are unknown in classic cooking and will not cause us regret over constructing such a large chart.

Consider, for example, that egg white beaten into stiff peaks, to which sugar and lemon juice are added, which we put in the microwave briefly; or the same mixture, which we bake like a meringue, and which results in a lemon "wind crystal" (a type of light meringue); or that mixture of beaten yolk and white, to which sugar is added (at this stage, we obtain the beginning of a preparation for genoise [sponge cake]), a liquid (try lemon juice), a little alcohol (rum? vodka?), and then microwave. (I suggest calling this preparation an Avogadro; it increases greatly in volume and is similar to a sabayon, poached rather than whipped while cooking.)

Let us stop there; conjuring this list reminds me too much of Garrigou's in the *Three Low Masses*, which leads to damnation. What becomes clear is that gourmandism may direct the course of invention!

Folding—A New Wrinkle in Cooking?

As we have known for four thousand years, making successive folds in dough exploits all the megalomania of the exponent.

As far back as the Neolithic, in making foods that require dough, bakers have used the process of folding. This repeated operation (think of puff pastry, or *mille-feuilles*: a thousand leaves) exploits the properties of the exponential, the royal way to very great numbers.

Cooks and then scientists have studied how the seeds of grains, crushed and then ground with water, form a dough in which the gluten proteins compose a network that imprisons the starch granules. Bakers roll this dough flat and then gradually work it thinner before baking it into creations like pastis, strudels, or croustades.

This family of dishes is distinguished from the crêpe, obtained beginning with a batter, which is then cooked on a hot solid surface. The starch granules swell, become "starched," bind together, while the excess water is eliminated; in that way we obtain a wheat or rice crêpe, a galette. . . .

FROM PUFF PASTRY TO NOODLE

To obtain the thousand-leaf puff pastry, beginning with a thousand leaves is not a good method, as we will discover if we try to repeatedly fold a sheet of paper in half. After six repetitions, the folding becomes difficult. However, bakers know that dough can be thinned when it is not yet baked. Thus leafed doughs are obtained by forming an envelope of dough, with a layer of butter in the middle. By rolling it out in one direction to make a rectangle, then folding it into thirds, and repeating this six times, we obtain 730 leaves of dough. With two layers of this kind we achieve the count of a thousand leaves, proof of honest advertising.

This invention dates back to when? Archaeologists at the University of Peking have shown that exponentiation was empirically employed in China four thousand years ago. The 2005 discovery by Houyuan Lu and his colleagues of noodles from late Neolithic China clarifies the history of the noodle. But who invented the noodle? Italy has long claimed to be its birthplace. An erroneous history of the food, a result of Italian pasta manufacturers' desire for publicity, credited Venice's Marco Polo with first bringing noodles from China. Food historians have countered those travel memoirs, published in 1299, with documents showing that pasta manufacturers already existed in the twelfth century. Thus it seems that noodles came from North Africa, passing through Sicily. Alsace, with its spaetzle, has also produced documents attesting to even older medieval noodles, so, until vestiges of Etruscan noodles are found. . . .

Italy? Alsace? North Africa? China? India? Technological analysis of the product leads us to believe that the invention of noodles must have very quickly followed the earliest use of grains and preparation of flour. How could the ancients not have invented strips for boiling?

A discovery made in China returns the credit for pasta (for how long?) to that country. Since 1999 Chinese archaeologists have been excavating the Lajia site in northwest China, along the Yellow River. Dated using carbon isotopes, this Neolithic encampment is found at a depth of 3 meters, under sediment. It was apparently destroyed following an earthquake that flooded the camp. Just recently, the archaeologists found an earthenware bowl sealed with a yellow brown clay, the contents of which were preserved. In the bowl they found noodles, as well as grain that had no doubt been used to make the noodles.

EXPONENTIAL ENTANGLEMENTS

Comparing this grain to grain from autochthonous plants, especially of the *Hordeum* (barley), *Triticum* (wheat), *Panicum* and *Setaria* (millets) genera, first led to the identification of *Panicum miliaceum* and *Setaria italica* as the species that provided the flour used for the Chinese noodles. Microscopic study of the starch granules from which the Neolithic noodles were made corroborated that hypothesis: The grain closely resembled that of millet.

The discovered pasta has a diameter of 3 millimeters and a length of 5o centimeters. How were they obtained? Exponentially! In fact, these noodles resemble the Cantonese noodles that today's cooks obtain by first forming a dough beginning with flour and water, then stretching the dough, folding it in half after having spun it in the air in order to expand it, folding it in half again after stretching it, and so on.

It is not difficult to calculate that, beginning with a piece of dough 5 centimeters in diameter and 20 centimeters long, it takes about ten operations to obtain noodles that are 1 millimeter in diameter. The most skillful Chinese cooks are said to be able to make their noodles pass through the eye of a needle. An easy mathematical feat . . . but a difficult technical one!

Let Us Eliminate Those Lumps!

How can we avoid the lumps that float to the surface in sauces? By breaking them up with water, which penetrates into the flour granules through percolation.

Let us pour flour very slowly into hot water: It falls to the bottom, proof that it is denser than water. Then let us pour a spoonful of the same flour all at once into a glass of hot water and observe: A mass forms and remains on the surface. It is a "lump." Why the lump? How to avoid it?

Let us analyze the problem. The flour that arrived in the water is composed of granules that are, for the most part, composed of concentric layers of two kinds of molecules, amylose and amylopectin, two glucose polymers. When such granules fall into hot water, the amylose leaves the granules and dissolves in the water, whereas the water enters between the amylopectin molecules and remains trapped there, which makes the granules float. These latter merge and form a "starch," as in a white sauce. This starch blocks the penetration of the water into the lump, while the air trapped between the granules sustains the lump, which floats.

SLOWING THE FLOW

For the physicist, starch is a gel, since it is composed of water dispersed in a solid. The water is trapped there because hydrogen bonds are established between the water molecules and the hydrophilic groups (the –OH hydroxyl groups) of amylopectin, and for hydrodynamic reasons. Indeed, we can consider the water to be found in the "canals" of the gel.

Now the flow of a liquid in a canal varies as the fourth power of the diameter. Assuming the canal is a micrometer in size (in reality it is much less), we calculate a flow that is expressed in millionths of a millionth of a millionth of a millionth of a metric cube per second. Nothing, practically speaking. Accordingly, once a gel forms, water enters the lumps that remain only very slowly.

Can the cook avoid this? To that end, recipes, methods, clever tricks abound, producing varied results, but could physics offer its more radical and definitive contribution?

Let us analyze the problem by proposing a model for it. The flour granules that fall will be considered as the knots in a net. The flour is not connected when it is dry; it is connected when the granules stick together.

Knots in a connected network? The theory of percolation allows us to describe such systems, in various cases, according to general rules that apply in all cases. For example, the grains of coffee in a percolator form a network through which water circulates; nothing flows until a continuous course is established between the top and the bottom of the network. The moment when such a course is established—and where the water flows, which is important for the coffeemaker—is the "percolation transition."

The theory of percolation applies as well to epidemics. A population is an assembly of individuals (the knots in the net) who exchange a disease, such as the flu; the transmission of a virus is like the establishment of a bond between knots.

SANDING AND PERCOLATION

The same principle applies for wire netting between the two sides of which is established a difference of potential; in this case the bonds between the knots are initially established, but let us begin by cutting some of them at random. As long as a continuous path remains between the top and the side of the netting, the current will pass; after a certain (statistical) number of cut bonds, the current no longer passes. The precise moment at which the current no longer passes is the "percolation transition."

Physicists have thoroughly studied this type of system and determined the percolation transitions, which depend upon the type of network under consideration. The exact value will not be useful in cooking, but the idea remains: If we see to it that a certain proportion of the bonds between flour granules are no longer established, the lump can no longer form. How to avoid the bonds between the granules? For example, by inserting, between the flour granules, granules of another material that does not jell, such as sugar, glucose, tartaric acid, or salt.

This experiment is conclusive. In a bowl let us mix flour and sugar (about one third), then let us take a spoonful of this mixture and pour it all at once into a glass of hot water: this time, all the powder falls to the bottom of the glass. The lump is avoided, thanks to a fundamental principle in physics.

Succulent Pearls

In mastering the laws of physical chemistry, we are learning how to make artificial caviar and many other strange gels.

In cooking, the classic savory gel is the one obtained from gelatin; cookbooks recommend the extract of calf foot, cooked in a bouillon that is then clarified (by adding an egg white that, in coagulating, traps the particles that cloud the solution). In addition to gelatin, could we not use jelling agents common in the food industry? The mad cow crisis prompted the (erroneous) fear of the presence of pathogenic prions in beef or pork gelatins, and cooks took an interest in fish gelatins, and then other jelling agents falling into the "terrible" category of food additives (although cooks have been using them since time began), among them, the alginates, extracts from brown algae. These molecules are polysaccharides, like starch or pectin, that is to say, long chains of sugars, in this case mannuronic and guluronic acids (molecules with many –OH hydroxyl groups and one –COOH carboxylic acid group).

JELLING AGENTS FROM ALGAE

Today the alginates, in vogue in top restaurants, allow various liquids to be trapped. Alginates are dissolved in a flavored solution, then drops of this solution are made to fall into a bath containing calcium ions. The alginates jell on the surface of the drops and trap the solution in a jelled film, its thickness depending upon the jelling process. That is how the food industry produced "artificial caviar" a few years ago, and modern cooks are producing what I have proposed calling "pearls": bouillon pearls (intensely flavored bouillon trapped in a thin envelope of jelled alginates), pearls of lobster bisque, pearls of passion fruit juice. . . .

Certain cooks have sometimes been bothered by the slightly rubbery texture of the drops they obtain. The problem is not difficult to control, because the various alginates offer different proportions of their two types of acids and, as a consequence, act differently as jelling agents. Different chemical compositions for the alginates lead to gels of different textures. Thus, alginates containing a high proportion of guluronic acid produce rigid gels, whereas alginates containing more mannuronic acid produce more elastic gels.

Another way of affecting the gels is by controlling the concentration of calcium ions that bind the acid groups. With high concentrations of calcium ions, the gels formed are very well bound, very firm, and resistant to heat. The acidity of the environment or the concentration of various other ions is also important; the solubility and the water retention of the alginates depend upon it.

In the 1990s the Danish physical chemist K. Draget proposed slowly releasing calcium in a uniform gel. In the journal *Lebensmittel Wissenschaft und Technologie*, Jong-Whan Rhim, of the University of Mokpo in Korea, studies the properties of alginate films formed in this way to better master the properties of such films and reduce the concentration of alginates.

FINE TUNING

J.-W. Rhim has studied two methods for producing these films: adding weak concentrations of calcium chloride to alginate solutions, on the one hand, and successive immersion of alginate solutions into solutions with increasing concentrations of calcium chloride, on the other hand. For all these gels J.-W. Rhim studied the mechanical resistance, elasticity, inflation in pure water, and various other physical properties of alginate gels.

Two mechanisms counter one another during the jelling process: the passage of alginates into solution and the reticulation by calcium ions. When the concentration of calcium ions is weak, the first phenomenon prevails, and the films are thin and supple; when the concentration of calcium ions is significant, thick, rigid films are obtained. The method that consists of using weak concentrations of calcium ions produces gels that are more fragile and stretch more easily.

J.-W. Rhim also considered the effect of a molecule that he dispersed in the gel and that does not participate in the latter: glycerin. The results obtained show that very different gels occur when the glycerin disrupts the gel and weakens it. Obviously, in cooking, glycerin must be replaced by a more comestible molecule, but cooks are used to experimenting with such plasticizers—that is how they use glucose to prevent table sugar (sucrose) syrups from crystallizing. Fats also can often serve to soften foods. There is a whole world to reflect upon in gels.

Tea Jellies

How can we obtain clear gels? By keeping the gelatin proteins from being precipitated by the polyphenols extracted from the tea during infusion.

In the past, only aristocrats clarified their bouillons, hoping, by doing so, to distinguish their consommé from the common gruel. Thus developed culinary techniques similar to the *collage* (gluing) of wines: With the help of egg whites, which are mixed into the hot bouillon, the particles responsible for turbidity are trapped, and, through filtration, a clear liquid is recovered.

Today, as heirs to the (political) ancien régime, we all want jellies, like bouillons and consommés, to be transparent.

ON THE ROAD TOWARD CONSISTENCY

How can we obtain clarity with tea, when the latter turns cloudy with the addition of gelatin? This question is very much in vogue, because cooks today are "deconstructionists": They retain the idea of classic dishes but change the organization and appearances. For example, for a beef stew they will serve the beef in the form of quenelles and the bouillon set in jelly, with puréed vegetables, and cornichons in the form of dried strips. Japanese cooks, *O tempura, o mores* (I insist upon the *u*), instead of serving sole meunière, might make fish with the texture of toast, or even as liquid as the butter, and give butter an appearance similar to that of fish.

Let us return to our tea, upon which we want to confer a jelled texture. Toward this end, let us make the tea through infusing tea leaves in hot water, and then let us add a sheet of gelatin—and it is a failure. The tea turns cloudy and does not jell!

An experiment makes the nature of this phenomenon clear. Replacing the gelatin, which is a protein, with an egg white, which is composed of 90 percent water and 10 percent proteins (ovalbumin, conalbumin, lysozyme, etc.), shows, by the cloudiness that results, that the proteins precipitate and are to blame. Acted upon by what? By something contained in the tea!

Let us examine our infusion again. Chemists have shown that first the odorant molecules of the tea leaves and then the tannins (in the case of tea, polyphenols) are extracted by water. Nothing surprising there. Black tea is obtained beginning with the immature stems of *Camellia sinensis*. Its principal constituents are the polysaccharides, like cellulose (about 20 percent), proteins (about 15 percent), varying amounts of polyphenols, caffeine (about 3 percent). During the tea's fermentation, which follows its harvesting, the polyphenols are transformed and give tea its specific properties, notably its color and astringency (the sensation of the drying of the mouth). The polyphenols combine with the proteins of the gelatin, which then aggregate to form particles that precipitate, which causes the cloudiness.

So how can we obtain a jelly from tea? An initial solution consists of treating the tea with a jelling agent that does not contain protein. For a long time the food industry has used polysaccharides in this way. Whereas proteins are chains of amino acids, polysaccharides are chains of simple sugars, immune to the action of the polyphenols. Tea does not become cloudy when they are used. Nevertheless, the texture and appearance of the gels obtained are not identical to those of gelatin gels. Alginate gels are rigid and opaque. Those obtained with a mixture of carrageenans and carob gum are elastic and clear, those made with xanthan gum and carob are elastic and opaque. . . . And we would like to retain the classic, pleasing texture of gelatin gels.

One solution consists of adding gelatin to the tea in a quantity just sufficient to make the polyphenols precipitate, and then either centrifuge the cloudy liquid that forms or let it sediment. The clear liquid recovered will be rid of its polyphenols so that it will gel without turning cloudy after a new addition of gelatin.

DISTINCT OPERATIONS

Such an exercise is difficult because the cook does not know the concentrations of polyphenols in the infusion, which can vary according to the nature of the tea leaves, the length of the infusion, the temperature of the water. . . . This solution presents the risk of the gelatin making the polyphenols precipitate, but also jelling the tea, which will prevent the sedimentation of the precipitate.

So why not "clarify" the tea with egg white and subsequently jell it with gelatin? Then nothing further remains except the delicate problem of the loss of odorant molecules over the course of the manipulations. But why couldn't cooks recuperate those molecules through distillations?

Gels, Gels, Everywhere Gels

Let us learn how to give gels a pleasing texture and good taste. Kientzheim sauce, an emulsion of melted butter in egg yolk, is a creamy delight.

We will now describe the invention of a butter sauce called Kientzheim, which takes on a supple consistency as it cools. Could this be a gel? Yes. But let us not harbor a prejudice against gels, sometimes pictured in the minds of consumers as those gelatinous, vibrating English desserts in various colors and with unlikely tastes that quiver under the tongue when you are lucky enough not to have them slide off the spoon on the perilous ascent from dish to mouth.

Without gel we could not survive: All around us everything is gel. Meat and fish are gels, as are vegetables and fruits, in which a continuous solid network encloses a solution. To these gels, with their separate aqueous compartments, the cook adds more connected gels, like the gel of gelatin, a network of gelatin molecules in which water molecules are trapped. In jams, which are also gels, it is the pectin extracted from the fruit during cooking that combines in a continuous network, when there is a sufficient concentration of sugar.

In addition to these physical, reversible gels, there are the irreversible, chemical gels. For example, cooking an egg white forms a gel made up of a network of coagulated proteins (the solid phase), in which the water (90 percent of the egg white) is trapped. Terrines and pâtés appear in the same category, since the proteins from the interior of the muscle fibers, released in the chopping, coagulate in the cooking. This effect is basic to the preparation of forcemeat, quenelles, croquettes, flans, set creams . . . in which the fatty substance in more or less significant concentrations is added to water dispersed in the solid. I was forgetting butter, in which a portion of the fatty substance forms a solid network in which the rest of the fat, in a liquid state, is dispersed (as well as a bit of water, emulsified in the fat).

GELS AND TASTES

Sapid or odorant molecules, present in solution, will penetrate a gelatin gel, for example. Those molecules injected into a gel diffuse at a rate that depends upon their size and the temperature and structure of the gel: A firmer gel, formed by molecules capable of binding to water molecules, will allow less rapid diffusion of water-soluble molecules, which, bound to water molecules, will temporarily bind to the gel. This peaceful invasion can easily be observed by injecting colored molecules, soluble in water (a syrup, for example): We will see that the diffusion is about a centimeter a day, which allows us to determine the time needed for a taste to reach the heart of a connected gel.

Let us come back to cooking by observing that the gel will be increasingly firm as the proportion of solid is greater, the solid is denser, and the dispersed liquid phase more viscous. For example, butter softens when the temperature increases, because the proportion of solid in it diminishes. What is the minimum concentration of solid necessary for the gel to hold? On the order of 1 percent, which explains why we can add water to the egg and obtain set creams nevertheless. Since a 60-gram egg contains about 7 grams of proteins, we can add nearly 700 grams of water and still obtain a chemical gel that holds.

A NEW SAUCE

And our recipe? Let us begin with an egg yolk, which contains 50 percent water, fats, and proteins. The addition of melted butter, drop by drop, to this yolk permits us, through dispersion (with the help of a whisk, for example), to obtain an emulsion of melted butter in the water of the yolk. The whisk divides the melted butter into fat droplets that the yolk proteins coat and stabilize.

What, by the way, is this sauce we obtain through the dispersion of melted butter in egg yolk? It is not a hollandaise sauce, since the egg yolk has not coagulated. It is related to mayonnaise, in which oil serves in the place of the melted butter. (Mustard has no business in mayonnaise, unless we are making a remoulade.)

Let us try a taste. The mouth first perceives the water in which the fat is dispersed, since this water makes up the continuous phase; then the melted butter comes to coat the mouth with an enveloping sensation. A taste of childhood, of home: This sauce we have conceived and christened Kientzheim, from the Alsatian *kind* (child) and *heim* (home). How can we enjoy it? Beginning with an egg yolk, whip the butter you use to cook your sole meunière into Kientzheim (while humming *O sole mio*). Happy holidays!

5

Understanding, Perfecting

Understanding, perfecting: Have we perhaps moved on to technology? Molecular gastronomy, which is a science, maintains a strange relationship with culinary technology, indeed even with technique. It feeds on the phenomena of cooking, a pedestrianism that is certainly no original sin, but, more important, the knowledge that it produces, for reasons difficult to understand, is immediately applicable in cooking . . . whereas science wants only to produce knowledge.

This case is not unique. Pierre Potier, the father of Taxotère (used in fighting breast cancer) and other anticancer compounds, made a completely exceptional and admirable specialty of studying natural products, specifically those coming from plants . . . which led to numerous molecules of therapeutic interest. Before him the great French chemist Michel-Eugène Chevreul, who lifted a corner of the huge veil by discovering the chemical composition of fats (triglycerides), contributed to a revolution in the making of candles. Examples abound that show how chemistry, physics (let us consider transistors), and biology (molecular biology in particular . . . but is that really so different from molecular chemistry?) can move almost instantaneously from the search for mechanisms to applications.

In the section that follows, it is a matter of cooking, but not of marginal cooking. I mean, it is not a matter of a few hard-boiled eggs with mayonnaise that could be eaten as a first course, or a few cornichons in vinegar to accompany a beef stew. No, this time, we are at the heart of the culinary question, and also at the heart of the scientific question. It is a question of the color of the dishes, the cooking of meat and vegetables, the creation of sauces.

Science considers a mountain and wonders, why did it rise here, and how? Molecular gastronomy asks the same questions; those hundreds of classic sauces are mountains, with compositions that need to be explored.

And that is its goal: to contribute to the intelligibility of the (culinary) world.

Prehistoric Cooking

Cooking with the aid of heated stones is possible, and the heated foods can be analyzed by the traces they leave in the rocks.

How did our prehistoric ancestors cook their foods? A number of archaeologists think that they cooked certain foods by boiling them in water where they deposited stones heated by fire. At the University of Rennes, Ramiro March and Alexandre Lucquin are exploring the "boiled-by-heating-stones" hypothesis by studying the food residues left on stones used for heating the water, or alterations in the stones themselves.

Boiling by heating stones was still practiced in North America when the Europeans arrived there, and in Ireland the technique survived until the beginning of the century. Nevertheless, the question remains: Was this technique used in the Upper Paleolithic? For half a century, many archaeologists have been reproducing the operation in order to study its parameters and to learn what traces they must look for at the ancient sites in order to verify the use of the technique.

In 1954 the Irish archaeologist M. O'Kelly was the first to study combustion structures at five ancient sites, to reproduce them, and to demonstrate experimentally that meat can be made to boil in the combustion pits of these sites. In a 454-liter pit he made water boil with the help of rocks present at the ancient site. The water boiled in half an hour. Then O'Kelly cooked a leg of mutton (eaten by the research team). Measuring the volume and the number of stones that could have been used for cooking meat, he calculated the number of meals that might have been cooked at the site and determined that, if this was the cooking process used by the site's occupants, they had stayed there for forty-five days. However, his reasoning was based on reusing stones without alteration.

In 1968 M. L. Ryder, inspired by the recipe for haggis, boiled meat in sheep's stomachs. The stomachs were heated by a fire situated 20 centimeters below them, and in some of them, he deposited stones heated in the fire. With this procedure, the stones made the cooking water overflow. The successive addition of stones that cannot be removed limits the volume that can be heated, and much water is lost. M. L. Ryder did not demonstrate that this technique had been used.

YOU ARE ROCK

In 1992 the Greek archaeologist P. Pagoulatos studied the thermal alterations in rocks and the possibilities for reutilization. These alterations are of three types: color, fissuring, and fracturing. The study was not comparative. To learn about the possible reutilization of rocks used for boiling, rocks heated many times and cooled in the air would have to be compared to rocks used many times for boiling.

R. March and A. Lucquin engaged in this study at the same time that they elucidated the conditions for cooking with heating stones. The first experiments were carried out on the feasibility of the technique: length of time for heating the rocks, volume of the rocks, temperature of the water, nature of the primary materials. They observed first of all that stones heat rapidly. In about 10 minutes stones placed in a wood fire reach a temperature of 600°C. The effectiveness of the process does not depend upon the nature of the stones, but the resistance to alteration differs widely according to the rock. Finally, the Rennes archaeologists observed that prehistoric cooks had to carefully calculate the addition of stones to avoid excessive boiling as well as the loss of water.

Next, the researchers wanted to learn if it was possible to recognize, through cooking residues left on the rocks, the type of foods that had been consumed by our prehistoric ancestors. Their first study focused on spinach cooked with the help of Fontainebleau sandstone. Under the accepted conditions cooking lasted more than 2 hours; then the rock was analyzed. Scanning electron microscopy revealed particular alterations in the reused rock samples: The cement that binds the quartz grains in the sandstone is broken down by the rapid variations in temperature during the immersion of the stones, and the quartz grains are altered. Additionally, the color of the rocks is modi-fied by cooking (whereas it is not by heating alone, with cooling by air or in water). This change seems to result from coloring molecules, organic in origin, present in the cooking water. The greenish coloration in particular apparently corresponds to the remains of leaves cooked and stuck to the surface. Also, carbonization takes place in contact with the hot areas of the rocks.

Do such modifications reveal the foods cooked in Paleolithic times? Oxi-dation of the organic compounds on the rock surfaces would have undoubt-edly destroyed any traces, but the researchers observed that the color at the heart of the rocks changes as well, because of migration of organic compounds through capillary action: All hope is not lost. Moreover, the residues from the degradation of those molecules would be markers of prehistoric cooking.

Lipids in particular are molecules that interest paleochemists because they are modified in a characteristic manner by cooking. (These lipids con-stitute 20 to 30 percent of the dry mass of spinach.) Analysis of stones used for cooking spinach revealed that these molecules are degraded by cooking, releasing chemically inert lipids, and their degradation has been much stud-ied (in petrochemical centers). The Rennes archaeologists observed a high proportion of the undecane molecule on the rocks, which would be a good marker of cooking.

Will chemical analysis soon reveal what people ate many hundreds of thousands of years ago?

Tenderizing Practices

Freezing squid, like freezing beef, tenderizes its flesh.

The milky flesh of squid inspires anxiety in the cook: How can it be kept from becoming exceedingly tough in the cooking? Consequently, cooks some-times indulge in doubtful practices. Some suggest tenderizing the squid, by cooking it in water with a cork floating in it, or by pressing it between two wooden boards in a vise, or by soaking it in carbonated drinks, or by freezing it for 2 hours and then pounding it on the work surface.

The last practice seems promising. At the University of Yamaguchi M. Ando and M. Miyoshi have explored the effects of freezing and of quick-freezing. Japanese interest in tender squid is not surprising: It is the most widely consumed invertebrate in the country.

CRYSTALS IN THE CELLS

We know that freezing and quick-freezing food alters its consistency. In 1988, at the Clermont-Ferrand INRA Center, J.-D. Daudin studied the formation of ice crystals in the flesh of animals. This flesh is muscle tissue, made of muscle fibers gathered into bundles by a tissue composed of collagen. The muscle fibers are cells measuring 10 to 100 micrometers in diameter, and many centimeters in length. The collagen is principally responsible for the toughness of the flesh.

When meat is cooled to temperatures below the solidification of water ($0\,°C$), ice crystals appear first between the cells, then within their interiors. Why? Because molecules in solution lower the freezing temperature of a solution. This cryometric reduction is what happens in automobile radiators when antifreeze limits the aggregation of water molecules into ice crystals. Since the concentration in solution is lower in the intercellular fluids, this reduction is less significant and crystals appear first between the cells. To this phenomenon that occurs during the freezing of products, another is added: the development of the crystals. Small crystals gradually disappear to be replaced by big ones. This destruction of bonds between fibers increases tenderness.

JAPANESE SQUID, LIKE BEEF

Does the flesh of squid behave like beef? In 1994 M. Kugino and K. Kugino demonstrated that freezing squid improves its tenderness. When cooked, frozen squid becomes more tender than nonfrozen squid. They also observed that the effect of freezing was greater on uncooked flesh than on cooked flesh.

Shortly afterward, American researchers D. Stanley and H. Hultin showed that this observation was only valid for Japanese squid *Todarodes pacificus* and that the flesh of frozen Atlantic squid became tougher! To explain this phenomenon the formation of formaldehyde in the frozen flesh was mentioned, since formaldehyde might strengthen the bonds between the protein molecules and toughen the flesh.

The conclusion is that the conditions for freezing are determining factors. Recently, M. Ando and M. Miyoshi have resumed studying the toughness of common Japanese squid that they froze one day after the squid were caught. The studies show that tenderization takes place especially during the first three days of freezing and that squid is not so different from fish: Freezing makes the same type of mechanical behavior appear.

DAMAGED CELLS

The Japanese researchers did not limit themselves to measuring mechanical characteristics; they also analyzed the liquid that oozes from frozen flesh and showed that the quantity of this liquid increases notably, not only during the first few days but also during the following month. This phenomenon no doubt results from the gradual damage to muscle fibers caused by the ice crystals: This damage would let muscle fibers "escape."

Finally, analysis of the structure through a microscope showed that the muscle fibers gradually separate when freezing is prolonged. Ice crystals form between the cells. They separate them and destructure the flesh; their growth, finally, leads to cellular damage that lets liquids ooze out.

Thus, the recipe that recommends freezing squid in order to beat it has its virtues. Let us note that the practice of beating to obtain tendernesses — as opposed to tenderness — has its disciples and its converts in other domains.

Five Per Day!

A few reflections on eating vegetables with pleasure by giving them flavor.

The specialists maintain that we should consume at least five vegetables and fruits per day to lower the risk of cancer and cardiovascular diseases. Gladly, but how should we prepare those vegetables? Contrary to the claims of bucolic Lamartinians, smitten with idyllic nature, it is not true that dishes are good when they have the taste of what they are. No, cooking does its best to transform the taste of foods, and the cook decontaminates food, changes its texture (either toughening or tenderizing it), and provides it with taste.

Without this decontamination we would be infested with the pork parasites or with liver flukes present on watercress. Washing eliminates some of the microorganisms that contaminate foods; paring removes bitter or toxic parts. Potato skin, for example, contains solanine, a toxic alkaloid that is dangerous when consumed in large quantities.

SLOW CONDUCTION FOR INSULATING MATERIALS

During cooking, microorganisms brought to a sufficiently high temperature for long enough a time are killed. This heating can be done through conduction, by placing the vegetables on a solid (in a saucepan or heated receptacle); in a liquid (aqueous solution or oil); or in an enclosure containing a hot gas (oven). Since vegetables are composed mostly of water, conduction is slow, and there is a risk of burning the vegetable in contact with the hot solid before the rest is sufficiently heated. Hot fluids, on the other hand, permeate all the crevices: hot air, boiling water, hot oil. This heating also deactivates the enzymes that would produce unappetizing colors. Water also eliminates various acrid or astringent possibly toxic molecules, and it deactivates molecules like the lectins, anticoagulants dangerously present in green beans, broad beans, and lentils.

WEAKENED CELL WALLS

Consistency is altered by physical and chemical actions. First of all, a hard vegetable mass can be divided up and rendered more easily assimilable. Since each plant cell is enclosed by a resistant cell wall, made (essentially) of cellulose and pectin, there is much advantage in attacking that wall and, especially, in breaking up the intercellular cement. Heat acts in this way, but chemistry offers other possibilities as well. For example, twenty-first-century cooks can use enzymes called pectinases, which degrade pectins. They can also add a basic compound (sodium bicarbonate, for example) to the cooking water in order to make the pectin molecules' carboxylic acid groups (–COOH) lose their hydrogen atoms, giving them the electrically charged form (–COO⁻), which results in repulsions that favor the tenderizing of vegetables. They can also use calcium complexants. This divalent ion links the pectins; by capturing it such bridging can be avoided and the vegetable tenderized.

Cooks can also prompt osmosis, which softens vegetables. Soaking a cucumber in salt makes it lose its crispness, as does soaking it in soy sauce, sugar, an acid (lemon or other fruit juice, vinegar, beer, wine . . .), just as alcohol modifies turgescence.

THE FANTASY OF THE RAW

Finally, we face the essential problem of taste. Contrary to an idea too widely accepted, cooking with steam does not preserve the taste of foods; it is impossible to get the taste of raw vegetables except by crunching on them intact. Cutting them breaks the cells, which releases plant enzymes and their substrata, which they attack. The best a cook can do is cook the vegetables in an emulsion (cream or something else) containing fat and water. Water-soluble odorant or sapid molecules will be dissolved in the water, whereas water-insoluble odorant or sapid molecules will be dispersed in the fatty material.

Cooks will benefit from using what chemistry can offer: For example, they can use caramelization, which brings into play the sugars that vegetables contain in notable proportions. They only need to heat the vegetable juices to eliminate the water and bring the vegetables to a temperature high enough to cause the caramelization of table sugar, sucrose, and the other sugars present (glucose, fructose, or inulin in onions, endive, asparagus, etc.). The Maillard reactions, on the other hand, will result if the cook adds the necessary amino acids by soaking the vegetables, before cooking them, in a solution concentrated in gelatin, egg white, fish or meat juices; then the vegetables should be heated as if to "lacquer" them. I call this creation "vegetable demi-glazes."

Numerous possibilities offer new tastes: Try tossing a few carrots in a centrifuge, collect the thick juice and heat it gently for a long time; that is how the French cook Pierre Gagnaire obtains an astonishing, delectable result.

The Green of Beans

The metals used in manufacturing cooking utensils help turn beans very green; thus they appear quite fresh.

The greenness of green beans is appetizing. How do we preserve it? Tricks and techniques, untested by science, abound on this subject. So it was that in 1896 the Paris cook Paul Friand wrote: "To retain the green tint of green beans, one must be very careful not to cover the saucepan. A tiny bit of bicarbonate added at the same time as the beans preserves the green color." In her 1925 best seller Madame E. Saint-Ange perpetuates another tradition: "If one wants to retain the very green tint of green beans, it is necessary to use, as great cooks do, a non-tin-plated copper utensil. Tin decomposes the chemical constituent of the green color."

Empirical ideas about the effects of acids and metals are still being circulated. Paul Bocuse writes: "To maintain their green color, a copper receptacle should be used if possible, this metal having the property of reviving the chlorophyll." Alain Ducasse's advice is "not to mix the beans with the vinaigrette in advance; the vinaigrette would alter their color."

Chemists at the Nestlé Research Center in Vers-chez-les-Blancs have perfected a method for analyzing alterations in chlorophylls and their derivatives during vegetable transformations; they have identified the effect of certain metallic salts on vegetable color.

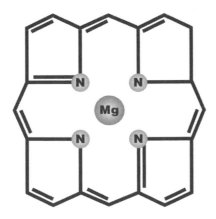

The color green in vegetables is due especially to the chlorophyll molecules in plant cells. When white light strikes beans, the chlorophyll molecules absorb certain visible rays, producing the color green. Chlorophylls owe their light-absorbing properties to their chemical structure: At the center of what is called a porphyrin group, four nitrogen atoms surround a magnesium atom. These nitrogen atoms belong to a group of hydrocarbonated rings, as in hemoglobin. (In hemoglobin, which makes the blood red, the center of the molecule is occupied by iron rather than magnesium.)

The chemist spots the possibilities for chemical reactions that would change the color of such molecules. In an acid environment the central magnesium atom is easily replaced by hydrogen atoms. That is what happens when beans are cooked in the presence of an acid and, no doubt, when they soak too long in a vinaigrette. The chlorophylls are transformed into a compound called pheophytin, which gives green beans an unappetizing yellow-brown color. Adding bicarbonate, thus rendering the solution basic (that is, less concentrated in hydrogen ions), avoids this yellowing.

That same magnesium atom can be replaced by other metals, as cooks who used to use "regreening tubs" have observed; the copper that replaced the magnesium gave the green beans a "fresh" green color. In the nineteenth century the food industry used copper sulfate to avoid yellowing in the canning process, but the treatment was forbidden because of copper's toxicity.

The degradation of chlorophylls in vegetables during cooking is a food industry problem. Since consumers judge the freshness of vegetables by their color, many research teams have endeavored to study the stability of chlorophyll molecules and have found other adjuvants besides copper: Iron and tin give a gray-brown color, but zinc gives a beautiful green color. Thus, according to the Veri-Green patent, green beans are first "blanched"; that is, heated briefly to inhibit the enzymes that degrade the chlorophylls. Then they are cooked in the presence of zinc salts. The effectiveness of the process has been attributed to the formation of zinc complexes more resistant to acids and to heat than the magnesium complexes.

The advent of modern methods of analysis, like high-performance liquid chromatography paired with mass spectrometry, has aided the study of these complexes.

That is how A. Gauthier-Jacques and his colleagues at the Nestlé Research Center have studied spinach from which the pigments were extracted. The first stage of their analysis, chromatography in the liquid phase, is a refined version of that high school experiment that consists of crushing leaves and depositing a drop of the crushed matter at the bottom of a piece of filter paper, the lower part of which has been soaked in an organic solvent, such as naphtha. As it climbs, the solvent separates the different pigments because it conveys them at different speeds, depending on their size and their solubility in the solvent. With liquid chromatography, the principle is identical, but the products are carried into a column. Behind the chromatography column, mass spectrometry reveals the mass of the separated molecules.

Analysis has revealed more than twenty-five compounds derived from chlorophylls resulting from heating; in addition to chlorophylls, pheophytins appear, in which the magnesium atom is lost. The other compounds derive from the original ones through the loss of more or less significant parts of the molecule.

Analysis shows especially how zinc interacts with and stabilizes the various chlorophyll derivatives. Equipped with such a useful tool, chemists will be able to specify exactly what cooking conditions are optimal for green beans. As for cooks, they will know that they are right to avoid long cooking, tin, iron, and acids.

The Taste of Roast Chicken

This taste changes with the cooking temperature. If you want to store meat and later avoid that "reheated taste," cook it at a high temperature.

What could be simpler than roast chicken: You take a chicken, you put it on a spit and heat it intensely. The skin crisps, the fat melts, the juices drip, and the flesh takes on a remarkable taste. Yes, but what taste? Derek Byrne and his colleagues at the Royal Veterinary and Agronomic University of Denmark have shown that the taste of roast chicken depends very much upon the roasting temperature.

The Danish chemists initially became interested in the taste of reheated meat. After a period of cold storage, products that are reheated acquire what is called a "reheated" odor, studied in the late 1960s. The subject remained dormant until the 1980s, when it was then accurately attributed to the phe-nomenon of auto-oxidation in fats, otherwise known as rancidity.

TURNING RANCID: A CHAIN REACTION

In chemistry, this auto-oxidation reaction is a textbook case. It brings free radicals into play; that is, molecules that are reagent because one of their electrons is not paired. The formation of one of these radicals engenders new radicals, propagates the reaction, and leads to a rapid process of turning ran-cid, especially if it is catalyzed by iron, ubiquitous in cooking (e.g., saucepans and animal blood).

The fats that oxidize the most in meat are the phospholipids that make up the cell membranes. These molecules include insaturations; that is, parts where the carbon atoms of the molecular structure are bound by double bonds. These insaturations determine their oxidizability, which increases, for animal tissue, from lamb (least) to fish (most), with beef, pork, and poultry in between, in that order. In addition to the auto-oxidation of fats, there are the so-called Maillard reactions, which alter proteins, prompting the disappear-ance of meat smells to the benefit of grilling smells.

We have known since about 1993 that cooking at high temperatures (above 100°C) inhibits the appearance of the reheated taste, because the mel-anoidins, brown compounds formed during the Maillard reactions, have anti-oxidant properties. However, the mechanism for these effects remains poorly understood.

D. Byrne, W. Bredie, and M. Martens in Denmark, with David Mottram in Bristol, divided up the research work, simultaneously conducting sen-sorial and chemical analyses on meat (boneless, skinless poultry breasts) cooked at various temperatures (160°C, 170°C, 180°C, 190°C), stored for more or less time (between one and four days), and then recooked for a precise amount of time at 140°C. To analyze odors the examiners established a list of descriptors: cardboard, linseed oil, rubber/sulfur, odor of chicken, roast, rancidity, vegetable oil, walnut. . . . For tastes they confined themselves to sweet, sour, bitter, salty, umami, metallic. At the same time, the products were analyzed chemically.

THE GUILTY PARTIES IDENTIFIED

The correlations of sensations with the experimental parameters showed that the odors of cardboard, linseed oil, rubber/sulfur, and rancidity increase with the length of cold storage, but the meat cooked at the lowest temperatures were most defective in this way. The odor of meat and the taste of meat vary conversely. The odor and taste of grilling increase with the cooking temperature, but—and this is a discovery that will most interest cooks in particular—different tastes result from different cooking temperatures. For example, meats are bitterest and most astringent when they are cooked at a high temperature.

The compounds responsible for these tastes and odors were identified. Most number among chemistry's respected and prosperous citizens. They were listed among the products of fat oxidation and Maillard reactions. As expected, the products of oxidation were found in high quantities for samples stored longest in the cold. The chemists noted the beneficial effect of molecules bearing a thiol group (a sulfur atom bound to a hydrogen atom). These molecules oxidize easily, thus acting like antioxidants, exactly like certain products of the Maillard reactions.

Research is now delving more deeply into the mechanisms of these effects, but already we have here a new refutation to a debatable "natural" trend that argues that raw foods might be good for our health. Not only does cooking kill the microorganisms present on the surface of meats, as well as parasites inside it, but it also combats oxidation . . . in addition to giving grilled meats a delicious taste.

A Small Exchange

The odorant molecules in cooking liquids permeate meat less than we thought.

We often cook meat in a flavorful liquid (daubes, ragouts, braises) hoping that the molecules imparting the flavor will penetrate the flesh. Cookbooks cite osmosis to explain the exchange of "principals" yielded by the cooking liquid and gained by the flesh. Will this optimistic hypothesis stand up?

Let us observe first of all that the theory is hardly good physics; if we liken the surface of the meat to a semipermeable membrane, the function of a concentrated cooking liquid would be to draw water from the meat rather than introducing the large, dissolved, odorant molecules into the flesh. Observation reinforces our doubts: Meat immersed in a cooking liquid contracts as the water inside the flesh is expelled into the liquid, which seems to preclude the introduction of odorant molecules other than by diffusion.

MEAT: A SPONGE THAT HEAT SQUEEZES

Why does meat contract? Because it is made of elongated cells, muscle fibers many dozen centimeters long. These fibers are sheathed in a hard tissue, made of collagen; collagen is a protein, with molecules that assemble spontaneously into triple helixes, which are organized into a "weave" — or rather a "mat" — made of joined proteins. This tissue, which sheaths individual muscle cells, also gathers them into fascicules, which are gathered into fascicles of a larger order, and so on, until they form a muscle. This collagenic tissue gives flesh its consistency and meat its toughness. (The flesh of fish, which contains little collagen, is tender.)

When meat first begins to cook, at between 55°C and 70°C, the collagen fibers contract and the meat shrinks to three-quarters of its initial length. The water contained in the meat is expelled, as if one were squeezing a sponge. Weighing the beef when one is preparing a beef bouillon will confirm this: The weight of the meat decreases with the loss of water. Now if the water is expelled, there is very little chance that the cooking liquid can enter the flesh. Nevertheless, when cooked for a long time, the collagen eventually dissolves into the water, forming gelatin, and contraction ceases. When the collagen is destructured, could the cooking liquid then enter the meat?

How can we test these hypotheses? A coloring agent is almost sufficient. If we place a cube of meat into a liquid in which we have dissolved a fluorescent colorant, and we follow the penetration of the coloring agent into the flesh. . . . After 20 hours of cooking, very little of the colorant has entered it.

It is possible that the colorant is in concentrations undetectable to the naked eye but higher than concentrations detectable to taste. Certain odorant molecules, in concentrations too weak for them to be identified chemically, have olfactory significance.

NOTHING ENTERS, EVERYTHING EXITS

Thus, precise amounts are necessary. Spectrofluorimetry, which detects with great sensitivity the presence of colorants in meat, reveals that various meats behave differently when cooked, and the presence of the colorant depends upon the nature of the meat. That is explained, of course, by the fact that the collagen content varies. Some meats fall apart when cooked because the collagen is dissolved and the cooking liquid penetrates the flesh. The meat thus takes on the taste of the liquid.

On the other hand, for meats such as the shoulder of beef, which does not fall apart in cooking, spectrofluorimetry detects no fluorescent colorant a few millimeters below the meat surface, even after many dozen hours of cooking. For these meats, it is useless to try to flavor them internally by cooking them in a flavored liquid.

One solution? Two, rather. First, a syringe will be very effective and fast for injecting the molecules into the core of the meat. Second, the odorant molecules of bouillon will better penetrate meat cut into thin strips, as for Chinese fondue.

Poet, Take Up Your Lute

Sealing casseroles with a lute does not prevent evaporation and the loss of aromas if the temperature of the liquid is close to the boiling point.

Black-legged chicken in a lute-sealed casserole. . . . The chickens are prepared and put into a casserole with an aromatic garnish; then the lid placed on the casserole is soldered to it with a cord of dough made of flour and water. Cooks imagine that the aromas are recycled in the sealed space of the pot, in such a way that they come to permeate the meat. Then, when it is served, when the maître d' breaks the hardened cord of dough, the guests are bathed in a flood of odors. The lute provides a grand spectacle, but are the odorant molecules really trapped? Let us keep the question of how well they permeate the meat for another time, and let us determine here and now if the lute seal effectively retains the odorant molecules.

"Wisdom," says the philosopher Alain Badiou, "calls into question things known and accepted, and is skeptical on principle, with a view toward becoming more certain." A pox on the dream of cooks and gourmands, let us question the effectiveness of the lute and let us experiment. The simplest test consists of comparing two identical casseroles, sealed with a lute or not. In the laboratory we can replace the casseroles with graduated glass beakers and the lids with watch glasses or with coffee cup saucers. On one of the beakers let us place just the lid. Let us "lute" the other; that is, let us solder the lid to the beaker with a thick cord of dough made of flour worked with water. In the same trial let us compare these two beakers to a third one left open. In the three beakers let us pour the same volume of water. Then let us put the three beakers into an oven heated to 180°C (a standard temperature for cooking chicken).

Let us then follow the variations in water level in the beakers over the course of time. We can see that the water evaporates more quickly in the uncovered beaker . . . but no difference appears between the beakers with the unsealed and lute-sealed lids. Could the lute be a useless culinary ornament? Cooks confronted with the results of this experiment that calls their knowledge into question argued, criticized the experiment, reproached the beakers for not being casseroles . . . but only those well versed in arcane techniques pointed out that the experiment as performed was worthless because the conditions for using the lute had not been respected; cookbooks do not say this, but oral transmission affirms that the lute is useful only over long cooking times, at a moderate temperature.

How moderate? Let us give this argument the benefit of the doubt and redo the experiment, placing the same three beakers in a 110°C oven. This time the water evaporates differently in the beaker with just a lid and the lute-sealed beaker; the lute-sealed beaker retains all its water, even after 4 hours of cooking. The culinary tradition is confirmed: If the lute retains the water vapor, it also retains the odorant molecules that would have escaped with that vapor.

How can we corroborate these results? If the water is retained in the lute-sealed beaker, then as it evaporates, the interior of the beaker must be under pressure. Let us measure the pressure by modifying the system a bit.

Even the most amateur chemist can use a small spirit lamp to bend a glass tube into a Z. On its side this tube is immersed into the water in a beaker, and the U of the tube that extends out of the beaker is then filled with a liquid (which does not evaporate: oil, for example). The lid is placed on the beaker fitted out with its tube and then sealed with a lute. The whole thing is placed in an oven held at the temperature of $110\,^{\circ}$C. The liquid in the U of the tube is thus under pressure from the vapor formed; in this way we can measure the pressure in the lute-sealed beaker and see that the pressure increases gradually in the course of the cooking process.

Science is not the accumulation of measurements but the exploration of mechanisms. How does the lute confine the water vapor? Many kinds of lutes exist, but the most commonly recommended are the most rudimentary: They are made of flour and water. The working of the dough prompts the formation of a network of proteins, the gluten, which traps the starch granules; thus the dough becomes "viscoelastic," that is to say, a bit viscous (it flows with difficulty) and a bit elastic (because of the gluten network). Heated, this dough dries on the surface, while the starch absorbs the water and becomes starchy; the starch granules bind together.

When the dough dries rapidly, as when the oven temperature is $180\,^{\circ}$C, it contracts on the surface (it dries), while the interior remains moist. Cracks appear in the crust through which the water vapor formed in the beaker escapes. On the other hand, when it is heated gently, the drying is more homogeneous, and the whole cord of dough contracts without cracking. The vapor is retained.

The culinary conclusion? Sealing a casserole with a lute is a good method for retaining the vapor and the odorant molecules in dishes, but cookbooks would be advised to recommend its use only at temperatures not much higher than the boiling point of water, because the pressure increases rapidly with the internal temperature T, as the fourth power of the quotient $T/100$.

Providing Taste

Taste for . . . bouillon, a sauce, an egg, meat?

Odorant molecules are hydrophobic and volatile. So how do they come to flavor the water in bouillons, juices, and sauces? All cooking boils down to

this question: Since meat, fish, vegetables, fruits, and eggs are mostly composed of water, how do we introduce odorant molecules into them to give
them taste?

We know that taste cannot be reduced to odor: The flavor is important, as
well as perceptions provided by the trigeminal nerve (which detects in particular the "coolness" of molecules like menthol in mint) and by the mechanical
and thermal sensors. Even if a bouillon does not contain odorant molecules,
it can have taste. But the fact remains that the olfactory component of taste is
significant. Eaters with a cold know this very well.

MINIMAL, BUT REAL, SOLUBILITY

Odorant molecules (as well as others) are not completely insoluble and
they can be characterized by a coefficient of division (log P) between two
phases, for example, between water and octanol (related to standard alcohol).
The two compounds form separate phases, even though one part of the water
mixes with the octanol (and vice versa). When a molecule is added to this system, it is divided between the two phases: The number log P is the logarithm
of the quotient of the concentration in the octanol by the concentration in
the water. Each molecule has its log P, positive if the molecule dissolves more
in the octanol than in the water, negative in the opposite case. Thus, vanillin,
which is divided 50 percent more in the octanol than in the water, has a log
P equal to 1.7.

Even a slightly hydrophobic molecule will dissolve to some extent in the
water of a bouillon, and since odorant molecules are often active in tiny concentrations, the effect will be perceptible. Better still, when the water is salted, the salt will reduce the solubility of the odorant molecule, which will help
to make it pass into the vapor phase. The eater will sense it more strongly.

THE TASTE OF WATER

Other methods exist for giving water taste. For example, if a more hydrophobic molecule were dissolved in oil, that oil could then be dispersed

in water to form an emulsion. Unfortunately, this emulsion is temporary, because various phenomena, especially a creaming of the fat droplets, make the oil rise again to the surface. Can it be stabilized? Readers of this book are invited to put their free time to good use by doing the following experiment with me. Let us dissolve a drop of oil in ethanol (standard alcohol), then let us pour this solution into water. The water will turn cloudy, just as it does when pastis is added to water. Initially, only the upper part of the solution is cloudy, but slowly the cloudiness spreads throughout the water. This cloudiness is a dispersion of oil droplets in the water. In place of oil, without great gustatory interest, we should be able to do the same experiment by first dissolving into the alcohol a solution of a molecule odorant in oil, or a pure odorant molecule that thus permeates the water thanks to the alcohol.

A CENTIMETER PER DAY

Getting odorant molecules into the water is a first step, which solves the problem of bouillons, stocks, and sauces. Could we follow this same reasoning to introduce odorant molecules into meat, fish, and vegetables?

Odorant and sapid molecules present in bouillons hardly penetrate meat because cooking squeezes meat like a sponge. On the other hand, immersing a hard-boiled egg in a solution of a fluorescent colorant soluble in water (and thus in egg, since the white is made up of 90 percent water) shows a slow penetration of colorant into the egg, exactly like the diffusion of a hydrosoluble colorant into a gelatin gel. Obviously, the more gelatin the gel contains, the slower the diffusion. We must also compare the two extremes, which would be, on the one hand, pure water, in which the diffusion is rapid, and on the other hand, pure gelatin, in which the diffusion does not take place. Cooks thus know how to provide flavor to foods through steeping: They only need to remember that the diffusion of sapid and odorant molecules will be limited to about a centimeter per day.

For cooks in a hurry there is recourse to a syringe filled with a drinkable liquid, which will introduce odorant molecules into the core of the foods. These "intrasauces," well studied in the 1920s by M. Gauducheau and widely used by those in the delicatessen trades, cut the Gordian knot of solubility.

Cardinalization

Why the shell of a lobster turns red when it is cooked and why the study of this reddening is of interest to cooks.

The lobster paraded on a leash by Gérard de Nerval was blue, because it was alive. The one for Salvador Dali's telephone was orange, because it was cooked. Michele Cianci and his colleagues at the University of Manchester have penetrated the mystery of this change in color and have explained why crayfish and lobster "cardinalize when cooked," as Grimod de la Reynière said in the eighteenth century.

In the early 1990s it was discovered that the red pigment in lobster, astaxanthin, takes on a blue color in the shells of uncooked lobsters because it is bound to a protein. It was assumed that cooking separated the two partners of this complex, releasing the pigment, which took on its natural red color again. But this explanation did not address the question of why the bonding of pigment to protein would alter the absorption of light. Chemists have answered this question. They have extracted the complex formed by the protein and pigment and have analyzed the interactions of the two partners using x-ray crystallography.

The astaxanthin in lobster shells is a carotenoid; it is related to the carotenoids that give carrots and tomatoes their color. At the center of the molecules appears a long chain of carbon atoms that are bound, one bond out of two, by a double bond. Accordingly, the electrons of the double bonds are "conjugated." Instead of being localized between two particular carbon atoms in the chain, they are shared along the whole length of the chain. Less bound

astaxanthin

to the carbon atoms, these electrons can absorb the photons (units of light) with weak energy, that is to say, with long wavelengths, in the red.

The chemists continued to wonder: Why would the astaxanthin and protein bond not have the reverse effect of changing blue into ultraviolet, or green, or yellow? We know that the rings of six carbon atoms at the ends of the central chain act upon the conjugated electrons and alter light absorption. When the rings and the central chain are situated in the same plane, the conjugated electrons can move even more freely. Does this effect take place when the astaxanthin is bound to a protein, in the shell, and does cooking (which disturbs the proteins) alter the color by changing the type of alignment? Questions abound and are all the more interesting since the light absorption of astaxanthin bound to a protein resembles that of retinal, which plays a part in human vision.

Many studies have gradually clarified the issue. First of all, it was discovered that the pigmenting complexes for lobster shells, the "crustacyanins," contain about sixteen molecules of astaxanthin and eight groups of two proteins. In these complexes, groups of two proteins (A_1 and A_3) link two molecules of astaxanthin through weak bonds.

Determining how free the electrons are in this type of structure is not to be solved by setting down an equation and resolving it. First of all, the actual geometry of the complex must be determined. The Manchester chemists crushed some lobster shells, dissolved crustacyanin complexes in solvents, and finally recovered crustacyanin complexes from the solutions formed. After the inevitable purifications, they formed crystals of the pure complex: four months of patient labor to obtain blue crystals measuring less than half a millimeter across. But their persistence was rewarded. The crystallographic studies by x-ray could begin, and they revealed the complex's geometry, which explains its color.

The change in the color of lobsters follows from the astaxanthin molecules being disturbed many times. First, the alignment of the rings at the extremities with the central chain increases the freedom of the electrons with double bonds. In cooking, the denaturation of each protein disturbs the complexes and leaves the extremities of the astaxanthin unaligned, which alters the absorption. Additionally, water molecules are released. As these molecules interact with the extremities of the astaxanthin, the color changes through dehydration.

COLORANTS THAT CAN BE MODULATED

What purpose do all these studies serve? First of all, they help us understand how lobsters avoid predators despite their naturally red, and thus easily spotted, pigments. Thanks to the astaxanthin and protein bond, the lobster shell takes on that dark blue color that ensures good camouflage in the depths of the ocean. Additionally, the studies by the Manchester chemists open new paths to food industry engineers. To obtain a variety of food colorants, they will be able to amuse themselves by sequestering astaxanthin molecules—a natural colorant and thus more valued by consumers than synthetic colorants—with proteins, using the mechanisms identified in the lobster shells. Finally, the cook will know that boiling shelled lobsters in a covered pot simultaneously provides proteins, which, cooked, produce powerful aromas, and carotenoids, which, heated, often produce aromatic molecules such as beta-ionone, responsible in part for the odor of strawberries and violets.

The Beautiful Captives (Odors)

When yogurt does not contain fats, adding thickeners will make it creamy. Unfortunately, this addition can inhibit the release of odorant molecules.

Low-fat yogurt, good for the figure, is mediocre for the palate. So the food industry adds polysaccharides to give it a pleasing consistency, which determines the overall taste of the food. At the Dijon INRA Station, S. Lubbers, N. Decourcelle, N. Vallet, and E. Guichard have studied how the addition of thickeners alters the odor of yogurt.

Thickening yogurt is like binding a sauce. Question: Why thicken? Answer: to stall the moment when the food is absorbed and increase the duration of the pleasure of tasting it. Nevertheless, this thickening has its downside because it slows the release of molecules with a gustatory effect: sapid molecules, which activate the taste bud receptors; odorant molecules, which stimulate the olfactory receptors by rising back up through the retronasal fossae in the back of the mouth; molecules that stimulate the trigeminal nerve, to produce the sensation of coolness or pungency.

TRAPPED ODORANT MOLECULES

Trapped odorant molecules produce effects only if they are released, and any interaction with other food compounds limits that release. For example, the amylose molecules of starch are long polymers that wind into helixes, forming cavities where odorant molecules come to lodge.

The amylose of starch is not an isolated case. This molecule, a chain of many glucose molecules, bears a very great number of (–OH) hydroxyl groups, as do all the polysaccharides, or complex sugars. Now the polysaccharides are used specifically as thickeners because the hydroxyl groups bind to water molecules; large molecules accompanied by a string of bound water molecules, they increase the viscosity from which results the thickening . . . and as a correlative, the unfortunate effects upon the odor of the foods.

One the one hand, the odorant molecules disengage less easily from a viscous solution; on the other hand, they interact with the hydrophobic parts of the polysaccharides. Depending on the case, either there is excessive retention and the food loses taste because it loses odor or the food remains longer in the mouth because the odorant molecules, weakly bonded, are released more slowly in the mouth. That is what cooks sense, but it is essential to do a precise study.

The physical chemists in Dijon examined the odor of low-fat yogurts. Yogurt is milk solidified through the action of lactic bacteria that, in consuming the lactose of the milk (the sugar naturally present in the milk), produces lactic acid, which precipitates the proteins from the milk. These proteins form a network that traps the water in the milk, as well as the fat that may be present.

The physical chemists studied strawberry yogurts to which thickeners had been added to the strawberry preparation that was mixed with the natural yogurt: modified corn starch, lemon pectin, guar gum, fructooligosaccharides. The strawberry preparations also contained aspartame and acesulfame potassium (sweeteners), fructose, calcium citrate, sodium citrate, strawberry pulp, and water. In sum, the strawberry yogurts were both viscous and elastic. The texture evolves a bit with storing because the lactic microorganisms continue to increase (slightly) the acidity, producing lactic acid in particular, which reinforces the milk gel.

WHICH THICKENERS FOR TASTE?

S. Lubbers and his colleagues then tested for the release of odorant molecules that had been added to the yogurts for the purposes of the study: four esters, the odor of strawberry, of fruit, of strawberry candy, and of flower, and one alcohol with the odor of green leaf. The odor of the yogurts was followed for twenty-eight days.

To measure the odor released by the various preparations that they studied, the chemists used a system very popular among odor specialists, which consists of immersing a polymeric fiber into the air that rises from an odorant preparation and then desorbing the molecules absorbed in a mechanism that identifies those molecules.

The chemists measured the release into the air of odorant molecules from their mixture, after its incorporation either into water, water with a fruit preparation added, or into low-fat yogurt. The odor diminished from the water (which retained few or none of the odorant molecules) to the water and fruit preparation and finally to the fruit yogurt. The thickeners reduce the release of the esters because, as predicted, the odorant molecules of the aromas become bound to the polymers these products contain.

Thus, let us thicken our foods with well-chosen thickeners, so that they can easily release their odorant molecules and their taste can be optimal. Bon appétit.

Twenty-three Types of Sauces

Classic French sauces number in the hundreds, but there are only twenty-three physical chemical kinds. Their classification allows for the invention of new ones.

It looks like a Prévert inventory: brown stock, white stock, poultry stock, game stock, fish stock, fish stock with red wine, essence of fish, various essences, glazes of poultry, game, fish, meat, brown roux, white. . . . Thus proceeds the litany of classic French sauces in the *Répertoire de la cuisine*, by Th. Gringoire and L. Saulnier (1901). How to find one's way? Physical chemical analysis has shown that all these sauces are of just twenty-three kinds. This classification allows for the creation of new sauces.

THE OLD CLASSIFICATIONS, THROUGH PRACTICE

The classic French sauces were classified by Marie-Antoine Carême, Jules Gouffé, Urbain Dubois. . . . One classification, in the classical spirit, was done by l'Académie des Gastronomes and l'Académie Culinaire de France in 1991. This classification is operational: Secondary sauces are derived from the mother sauces. For example, clear brown veal stock is prepared beginning with calf bone that is browned in the oven, then cooked for 2 hours with water, carrots, onions and garlic, tomatoes, and a bouquet garni. From this clear brown veal stock is derived bound veal sauce, by adding a roux to these juices, and then cooking it for a long time. From this bound veal stock can be derived African sauce, anise sauce, Bercy sauce, and so on, through the addition of elements giving a specific taste.

This classification describes operations, not results, in such a way that related sauces are nevertheless distanced from one another. For example, crème anglaise, obtained by cooking egg yolk, sugar, and milk, is not considered to be related to hollandaise sauce, obtained by cooking egg yolk, an infusion of shallots, and butter. Nevertheless, in these two sauces the viscosity results from the coagulation of the egg yolk and the emulsion of a fat.

A system that I introduced during the European Conference on Colloids and Interfaces in 2002 offers a new classification, based on the physical chemical structure of the sauce. In it, G indicates a gas, E an aqueous solution, H a fat in the liquid state, and S a solid. These "phases" can be dispersed (symbol /), mixed (symbol +), superimposed (symbol ø), included (symbol @). Thus, veal stock is a solution, which is designated E. Bound veal stock, composed of starch granules swelled by the water they have absorbed, dispersed in an aqueous solution, is thus described by the formula $(E/S)/E$.

For many sauces the physical chemical formula is complex, because the recipe calls for numerous ingredients. A simple ravigote sauce, made beginning with oil, vinegar, capers, parsley, chervil, tarragon, onions, and salt, has a formula in which each ingredient is represented. Nevertheless, an aggregate of a few vegetable cells, whether they come from onion, chervil, or parsley, remains a microgel, because the water from the cells is dispersed in a small solid. The structure of the sauce does not depend on the nature of the aqueous solution thus used, so that the formulas of the sauces are simplified. In sum, the hundreds of classic French sauces are reduced to twenty-three categories.

A BASIS FOR INVENTION

What to do with such results? First, observe that certain categories of formulas are missing. Why, for example, does no sauce have the formula $G+(E/S))/E$? Such a sauce is obtained by adding a stiffly beaten egg white to a velouté sauce, which is itself obtained by cooking a roux in a bouillon.

The knowledge of the final state of sauces allows us to find different means for obtaining them. For example, Michel Menant sauce is obtained by cooking shallots in butter, adding Noilly Prat, reducing, adding concentrated fish stock, reducing again, adding port, reducing again, adding cream, reducing again, filtering, and adding butter and thyme. The same result can be obtained by mixing all the ingredients at once and then reducing it all together. And the taste?

Finally, knowing the classic formulas allows us to obtain modern sauces in the classical spirit. The velouté sauce with the addition of stiffly beaten egg white is not classic because no classic sauce has that same formula, but a sauce in which the oil phase, obtained by melting butter, might be replaced by a melted foie gras, would be in the traditional spirit.

Of What Use?

A study of two classic operations: skimming sauces and flambéing. Are they really useful?

FLAMBÉING

Everyone knows about flambé. We flambé crêpes suzette, baked Alaska, some warm wines, and sauces that contain cognac. It is an odd operation. Why does heating a saucepan that contains cognac, without flambéing it, produce a different result from heating an identical saucepan and lighting the vapors of the evaporated alcohol? In both cases the same initial ingredients are present, the same heating takes place, and, in both cases, the same evaporation carries the same compounds out of the saucepan.

Does the flame above the saucepan that is being flambéed produce new molecules? Certainly, the combustion of ethanol vapor (and other organic compounds evaporated with the ethanol) forms new compounds, but the vapors and the flame that rise carry these compounds out of the saucepan . . . in such a way that the contents of the two saucepans ought to be identical.

Unless the flame itself heats the liquid below it so much that new molecules are formed in the solution! How much does the surface temperature of the heated liquid increase when the vapors are ignited? The results of the experiment that involves placing a thermocouple in such a flame and in the liquid over which it rises will not surprise anyone who understands the physics of the evaporation of mixtures: As long as the alcohol is present (what we see in the flame rising from the liquid), the temperature of the liquid remains the same, below 100°C, and most important, there is no difference in temperature between heated and flambéed cognac, and the same cognac heated in the same fashion, but for which the alcohol vapors are not ignited.

In the flame itself the results of the measurements are more surprising: If temperatures over 200°C are attained toward the top of the flame, the base remains at a constant temperature of only about 85°C, a few centimeters above the surface of the liquid, as long as the flame is present. How would such a low temperature alter the taste of the flambéed sauce?

Chemical analysis of flambéed or nonflambéed sauce has yet to be done, but let us wager that flambé will reveal itself to be a sham. Not in every case, however: Cooks who make crêpes suzette or baked Alaska are well aware that the irregularities in the surface of crêpes or beaten egg whites that are licked by the flames brown more than the other parts. In this case the taste is clearly altered.

SKIMMING

Skimming is a classic operation in French cooking, which, according to the *Larousse gastronomique*, aims at "removing all the impurities that, over the course of slow simmering, rise to the surface of a stock or a sauce, and form a disagreeable scum."

Let us imagine, for example, that we have prepared a velouté sauce, beginning by cooking butter and flour until golden brown (making what is called a roux) and then adding to this paste a fish or meat stock. Cooking the preparation thickens the sauce, because the starch granules from the flour release a portion of their amylose molecules (molecules formed from linear chains of glucose molecules), whereas water penetrates between the amylopectin molecules (branched glucose polymers) that remain in the starch granules. This is how the skimming occurs: The saucepan is placed on a very gentle flame, in such a way that a single convection cell makes the heated liquid rise from the bottom of the saucepan. At the top of this convection cell the impurities that accumulate are eliminated at regular intervals with the help of a spoon.

An interesting practice, but what are the impurities? Excess fatty matter? Particles that cloud the stock poured into the roux? Wanting to learn if the fat from the initial roux was gradually eliminated over the course of the skimming and if the skimmed sauce only retained fat bound to amylose, we prepared a velouté sauce model with the help of René Le Joncour and Raphaël Haumont. Butter, flour, and water (rather than bouillon) were assembled according to the rules of the art. Then, at regular intervals, we removed the skin that formed on the surface of the sauce, measuring the temperature at various points and observing the microscopic state of the sauce.

A delicious odor of mushroom soup arose after half an hour of cooking (even though only plain water had been added to the roux) and a skin formed, at the beginning, about every 12 minutes. We extended the skimming in order to be sure we removed all the impurities . . . but, after about 10 hours, the sauce had been so well skimmed that nothing remained in the saucepan! Other experiments confirmed these initial results, so that we now know that the sauce itself is what cooks take for impurities when they skim. Of course, some real impurities are eliminated at the beginning of the operation, but the skin that forms is not composed of such impurities; the sauce itself forms a skin on the surface.

And the fatty matter? It is less stabilized at the end of the skimming, but observations through the microscope show that it remains dispersed in the same fashion and in analogous proportions at the beginning and the end of the operation. From which the inevitable conclusion arises: Too thorough skimming, advocated by the church fathers and certain cooks, is harmful.

The Eaux-de-Sauce

Like eaux-de-vie, these novel preparations, often amber colored, result from separation and taste delicious. Try them!

Although they coat foods, sauces are most often liquids, with varied qualities of flow. Thus mayonnaise, which seems solid at rest, fluidifies in the mouth when it is put into motion. The art of making sauces is to obtain the characteristic behaviors of each sauce through binding them; for that, sauce makers use a protein, which they make coagulate (egg, blood), or they add fat, melted and dispersed into a myriad of droplets in the continuous aqueous phase (emulsions), or they use a variety of starches (rice, wheat, potato, etc.) that, heated in the water of the sauce, swell as they absorb that water.

In sum, the sauce maker seeks to obtain a preparation neither too liquid (a sauce is not a juice) nor too solid (a sauce is not a puree) that will coat the food morsels in the dish (meat, fish, vegetables).

FROM A SMALL ILL, MAKE A GREAT GOOD

A sauce that separates into many phases—we French say that it "slices"—is a failure, a mistake. And if we turned it into something desirable? For example, if we managed to retain only the aqueous part, would not the clear solution have virtues analogous to those of the loveliest consommés, those bouillons enriched by the taste of the meat that is cooked in them, then clarified with the help of an egg white?

As a test, let us begin with a stew sauce, obtained by cooking meat in wine, with onions, carrots, bouquet garni. . . . During the long cooking of the stew, the cooking liquid was first thickened by way of the flour used to dredge the meat before browning it in the oil, then fluidified, being enriched especially by the hydrolysis of the amylose and amylopectin in the flour. A dark, cloudy sauce is thus obtained, which normally would have been thickened by the addition of blood.

This time let us try filtration or clarification. Filtration is the simplest method, but sadly it remains confined to laboratories, with their sintered glass funnels with controlled porosity. Nevertheless, we may benefit from centuries of development in laboratory techniques. The catalog of laboratory materials offers filters of all kinds, even systems that do not clog up. Failing that, we can filter with makeshift systems, clean sand placed in a cloth, for example, the whole assembly placed in a cooking strainer.

For those not tempted by this system, there is still the classic clarification of bouillons, achieved by adding beaten egg white to the cloudy sauce, then cooking for a long time, before passing the liquid obtained through a cloth folded four times, lining the sides of a strainer. In both cases, a clear liquid is recovered, amber-colored, like a fine cognac . . . that we can serve, like cognac, in a glass on the side of the meat. Let us call this superb product *eau-de-civet*.

Let us extrapolate to the whole range of sauces. In 2003 we demonstrated that all the sauces in the *Répertoire général de cuisine*, by Th. Gringoire and L. Saulnier (1901), can be reduced to fourteen physical chemical types. Following these first results, the body of work to be studied was expanded to include two principal works on cooking: *La cuisine française au XIXᵉ siècle*, by Marie-Antoine Carême, and the *Guide culinaire*, by Auguste Escoffier. Today the number of physical chemical types of sauces has stabilized at twenty-three. Most of them contain a dispersant aqueous phase, in such a way that most classic sauces can give rise to *eaux-de-sauce*, analogous to *eau-de-civet*.

FILTRATION, DISTILLATION, . . . AND MORE

How can we produce the sauces? Filtration and clarification can be used. However, since sauces often contain a fatty phase, which clouds the eaux-de-sauce, adding fat should be avoided before filtration or clarification. If it cannot be avoided, we can try instead to destabilize the emulsions and skim the sauces to recuperate the fatty phase, which can be used as well.

Destabilize the emulsions? Chemists are in the habit of passing their emulsions over steel wool. We can thus envision using distillations in those cases when sauces form emulsions that are too stable. After all, if the process is forbidden for the production of alcohol, it remains legal for culinary operations.

You have no retort? Not a problem: Fit a rubber tube onto your pressure cooker, in place of the safety valve. In this system, the sauce will cook at atmospheric pressure, and you will recover from the outlet of the tube only the aqueous part of the sauce. (To make the vapor condense, immerse an elbow of tubing into cold water.) Have a taste: delicious!

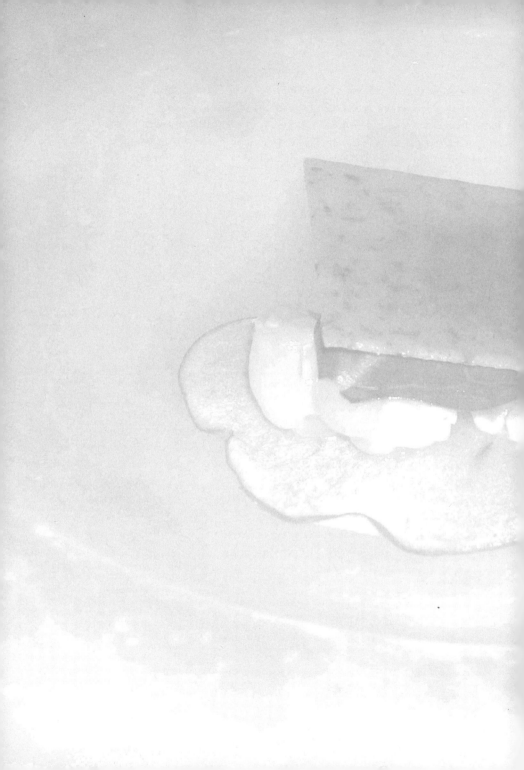

6

Without Forgetting All That Makes Life Beautiful

Yes, our world is a terrible place, and many of our human brothers and sisters die of hunger. Is not molecular gastronomy a kind of extravagant flourish? When Lavoisier studied beef bouillon in the prehistory of our discipline, he was concerned with making the best possible use of the meat allocated to the hospitals by the king. Here science was certainly not superfluous, and the work of the patron of chemists was, like that of many chemists of the past, concerned with the food of the populace.

Could molecular gastronomy be different in nature? Could it be aimed at higher delights, like those evoked by Brillat-Savarin? No indeed. Those—and especially those—who do not have the means to buy expensive meat must have the means to tenderize it, to increase its value. It is a vital question of domestic economy, an old-fashioned term that deserves to make a comeback on the culinary scene, by way of schools, especially. Yes, molecular gastronomy, insofar as it must be concerned with its applications, is not the prerogative of a wealthy class but, on the contrary, a science that produces new knowledge and gives everyone alike the means for transforming ingredients more effectively than tradition allows.

The question of the pleasure of eating remains, a pleasure condemned by a certain religion, because it does not elevate the soul. That requires the hair shirt, discipline, fasting. . . . Once again, the theory can be contested. Did not Saint Vincent de Paul say, "It is necessary for the body to be at ease in order for the soul to like it there"? Yes, such a declaration risks justifying all kinds of excesses, but on his side, Brillat-Savarin set limits to such excesses: "Men who stuff themselves and grow tipsy know neither how to eat nor how to drink."

Having established the framework, having been authorized to concentrate on culinary details by the moral authorities (ethical authorities, rather, the philosopher Hegel would say), we can now take an interest in all that makes what is good become better. The bubbles in champagne, the pastis in the aperitif, the varieties of cheeses, the freshness of bread, the meringues that offer their crispness to desserts, the bouquet of port wine. . . . Yes, we can survive on Spartan gruel, but why not enjoy life and the delights that our forebears perfected . . . or perfections that are incumbent upon us to discover?

Food . . . Earth

Clay containers alter dishes by contributing flavors that were once appreciated, or by eliminating undesirable molecules like tannins by trapping them.

Today only children, animals, and the sick eat earth. The first do this as part of exploring their environment; the second make the most out of clay, which, when combined with vegetable tannins, eliminates their astringency; and the third use this same "complexation" mechanism for curative ends. In the food history of France the use of earth is an important culinary factor that Danièle Alexandre-Bidon has studied.

In the twelfth century the Arab geographer Idrisi reported on a clay of Spanish origin with an agreeable taste, and *The Book of Wonders*, published by Gervais de Tilbury in the thirteenth century, mentions a field in Hebron from which inhabitants extract a red clay "that is eaten." Exported to Egypt, this clay cost as much as the most expensive spices because the odor of wet earth released by the clay pots containing various solutions was heavenly. Milk, honey, water took on a pleasing taste. According to *The Book of Foods and Health Preservation*, written in the thirteenth century by Ibn Halsun, springwater ought to come from a clay soil of good quality so that its taste will be "sweet."

Cooking in earthenware served the function of sweetening: "the dishes one sweetens in vessels of earth." What was this sweetening? A matter of adding sugar? Of eliminating the astringency of the tannins? Of providing characteristics appreciated in that diet?

SWEETENING . . . REALLY?

For those epochs gustatory significance must be interpreted in the light of the theory of humors, in which earth plays an essential role. The diet in those times was based on Aristotle's idea of the four elements: fire, air, water, earth. One sought to bring together in the diet substances that "balanced themselves" and to compensate for the possible excesses in the temperament of the eater. Considered to be cold and dry, the clays corrected excessive "humidities."

The chemical composition of clays and clay pots truly does prompt drying. Clays, which are oxides of various elements, such as silicon, are organized in layers to form particles that collect into aggregates. Their capacity to pump water can be considerable, and their exchanges are determined by the various types of pores. That is why, in hot countries, water is cooled in a porous jug. The evaporation on the outer surface draws energy from the water that is in the jug. Additionally, the crevices on the surface increase its area and the clay traps various molecules. This property has been put to use since at least the fifteenth century. The Florentine humanist Alberti mentions that in Toulouse people add clay to their wine in order to obtain "more flavor." Was it a matter of "improving" wines by adding taste to them or by removing tannins?

The *Mesnagier de Paris* (1393, one of the first cookbooks in French) specifies that "soups are better if they are made in earthenware pots." If the effect exists, is it due to a selective concentration of the bouillon, a portion of its water crossing through the clay, leaving the hydrophobic organic molecules (especially the odorant molecules) in solution? Or does the clay's weak thermal conductivity prevent excessive boiling, which would make the odorant molecules disappear, carried off by the water vapor? For meats, would clay's poor conductivity necessitate long cooking times, so that the tough collagen in meat would have time to be dissolved, while the slowly extracted gelatin was hydrolyzed, forming a solution of amino acids with a powerful taste? These are open questions.

EARTHY SCENTS

Chemists will also be able to verify Olivier de Serres's assertion that the "earthy scent" of clay vessels can be avoided by boiling them a long time with some bran. Does the bran capture odorant molecules absorbed in the clay? Is it true that a new pot, heated and then broken into pieces, gives a "potable" taste to wine or water? And does "potable" here mean that it is drinkable, or that "it has the taste of a pot"? Is it true that "bad wine spoils and corrupts the [earthen] vessel if it remains in it a long time"?

The cook has gradually eliminated interactions between the contents and the container; stainless steel has taken over and winegrowers are almost alone in using the virtues of wood to change the taste of wines. Why should cooks not select clays to improve the taste of dishes according to the ions that the clays provide or remove? By avoiding the odors of wet earth, we avoid the topsoil, with its many organic molecules resulting from plant decomposition. When will we use the powder provided by the crumbling of "earth poorly baked" in everything from pears to wine? When will restaurant menus really offer dishes with a true taste "of the land"?

Artichoke Cheeses

Using vegetable enzymes, it is possible to make cheeses without rennet. The discovery of an enzyme in the artichoke blossom that makes milk curdle confirms a bit of folk wisdom.

As early as 1655, in *Delights of the Country*, Nicolas de Bonnefons wrote: "If you take the violet beards of Spanish cardoon flowers and you let them dry like roses, you can use them in place of rennet to make milk curdle." Popular wisdom offers other milk curdlers: nettle leaves, blessed thistle seeds, wild thistle flowers, ginger, raw artichoke choke. . . . At the National University of Lujan in Argentina, Berta Llorente, Cristina Brutti, and Nestor Caffini have identified the milk-curdling enzyme that the artichoke contains.

Milk is a dispersion in water of fat droplets and protein "micelles," assemblies of proteins called caseins and ions, especially phosphate and calcium. The electrically charged caseins prevent the aggregation of casein micelles. When a rennet enzyme (chymosin) cuts the caseins that no longer play their part as a barrier, the milk coagulates. The biochemists found other coagulant enzymes for making cheeses, in particular extracts from cardoon flowers (as the traditions indicated).

Whether they come from the seeds, leaves, or flowers of various plants, all the enzymes used in modern cheesemaking are so-called proteinases (or proteases) of aspartic acid, having analogous amino acids sequences. Two groups of these proteinases have already been isolated from *Asteraceae* flowers (the cardoon family) and used by the food industry for making cheeses with a taste and consistency different from those made with milk curdled with rennet. These successes rekindled explorations of natural products, in particular the artichoke. If Jean Froc and his cheese specialist colleagues at INRA have long since confirmed the effect that cooks like Marc Meneau put to use, which molecules from which parts of the artichoke make milk curdle, and why?

COAGULANT EXTRACTS

The Argentine biochemists first prepared extracts of various parts of the artichoke through filtration and centrifugation of crushed tissue. The coagulant activity of the extracts was then tested on standardized milk samples, obtained by dissolving into water measured quantities of powdered milk, and the time between the addition of extracts and coagulation was measured. To test this action precisely, the biochemists also measured the activity on the caseins by mixing the extracts into solutions of these proteins. Finally, the quantity of the enzymes was measured. These tests showed that coagulant enzymes are present most significantly in the mature flowers, ten times more so than in the immature flowers and in the leaves (activity analogous to that of immature flowers).

Since the flowers are the only parts of the artichoke with coagulant activity, the study focused on them, and the biochemists eventually isolated a proteinase from the immature flowers. They purified it using chromatography, analyzed the amino acid sequence, and then compared that sequence to the proteinase sequence of other flowers.

CURDLING AGENTS: CARDOONS, ARTICHOKES

As with the cardoon proteinases, the pH at which the extracts of mature artichoke flowers are the most active is between 4.5 and 5.5. The use of various enzymes or compounds that reveal the nature of enzymes has shown that the active enzymes belong to aspartic acid proteinases, like the other proteinases isolated from *Asteraceae* flowers. Lastly, chromatography revealed five peaks, the last two of which were especially active; they corresponded to proteins with a molecular mass of about 60,000. Further analysis showed that the curdling proteinase in artichoke flowers is synthesized in the form of a proenzyme with a molecular mass of 62,000, which is then divided into chains of masses equal to 30,000 and 15,000 respectively, combined in the active enzyme.

The Argentine biochemists thus discovered a system analogous to the one that had been observed for the enzyme of the cardoon flowers, with a molecular mass of 64,000, which was then divided into two chains.

I forgot! If you want to make your cheese without rennet, the recipe is as follows: Cut the choke from the artichokes before cooking them (or else the enzymes will be denatured); place this choke in a muslin cloth and let it soak in the milk in a warm place for one night, in proportions of one choke for half a liter of milk. It is not against the rules to season it, in order to transform the experiment into a culinary success.

Worked Cheeses

A chemistry game allows us to extend the fondue to all kinds of cheeses. The results are better with matured cheeses.

Emulating the anonymous inventor of the fondue, cooks are trying to "work" cheeses. By heating them and mixing them into a liquid, for example, reblochon into carrot juice, they are trying to create a macroscopically homogeneous phase. They do not always meet with success; if they are not careful, they may find an elastic, rubbery mass swimming in a cloudy liquid. . . .

CALCIUM PHOSPHATE, THE CHEESE CEMENT

The behavior of cheeses derives from their composition. In the water component of milk, which cheese comes from, fat droplets, globules, and casein micelles are dispersed. These micelles are aggregates of many kinds of proteins cemented together by ions, phosphate, calcium, or citrate. The globules of fat are coated with a membrane, made of proteins and lipids, that ensures their dispersion in the water. Milk is at one and the same time a solution, a suspension, and an emulsion.

The curdling of milk leads to cheese. The addition of rennet, an enzyme extract preparation from the fourth stomach of the calf, modifies the milk's casein micelles, which aggregate. The mass that results from this aggregation still contains water, proteins, and fatty matter. Milk can also be curdled through the action of microorganisms that transform lactose into lactic acid. This acidification aggregates the micelles because the negative charges of the casein micelles are neutralized and no longer repel one another. Thus cheese principally contains aggregated caseins, water, and fat.

WHEN THE DEGRADATION OF A SUGAR FAVORS A PROCESS

During the maturing process for cheese, the lactose is transformed into lactic acid, the fatty acids are detached from the fats, and the caseins are broken down. Depending upon the cheese, 20 to 40 percent of the caseins are transformed into soluble proteins. Thus well-matured cheeses are easily dispersed in water, forming emulsions similar to fondue, whereas young cheeses will remain in an undesirable mass.

To obtain good dispersions offering smooth textures, the makers of spreadable cheeses heat the cheese in the presence of "melting salts," which contain lactate, citrate, or phosphate ions. Citric acid, for example, in binding with calcium ions, breaks the calcic bridges that link the submicelles into micelles, which disperses the cheese's proteins into the solution. However, the acidity of the citric acid has the opposite effect and prompts coagulation, so cheesemakers add citrate ions in the form of sodium citrate (E331).

AND WHEN CITRATE IS LACKING

Since amateur cooks do not yet have the citrate that would help them work cheeses, they can use fondue cheese or cheese spread, which contain precisely the same very effective products. More traditional cooks who fear UFMs (unidentified flying molecules) from food industry products can use more "culinary" means. For example, lemon juice contains citric acid, which sodium bicarbonate neutralizes, and the combination of the two allows for the dispersion of proteins.

There remains the problem of the dispersion of fatty matter in the aqueous phase, or emulsification. Toward this end the cook will have to use surface-active molecules. In a mature cheese these are abundant because the proteins have been dissociated, but in a young cheese, a food emulsifier is required. Gelatin is the ingredient of choice (though many others exist.)

The cook who has thus extrapolated from the recipe for fondue will be able to add a last refinement: avoiding the coalescence of fat droplets by increasing the viscosity of the aqueous phase. The food industry uses hydrocolloids, which increase the viscosity by binding to large quantities of water; to name a few: alginates, agar-agar, gum arabic, pectins. . . . More simply, a pinch of flour works wonders.

Cornichon Pickles

Let us use salt to avoid too much vinegar, zinc in place of copper for the color, and enzymes to maintain the crispness of cornichons.

The season is at its height, the cornichons are growing nicely. To have them available all year round, we have to cook them and pickle them. How do we go about it? The culinary tradition will not be much help, unless we interpret it through the lens of recent studies in biochemistry. In *La cuisine moderne*, published in 1885, a "gathering of cooks" reveals the characteristics desired by cornichon pickle lovers: They must be crunchy, not too vinegary, and green.

Not too much vinegar, first of all. That demand is paradoxical because the acidity of the vinegar is indispensable. It prevents the colonization of microorganisms that would degrade the plant tissues. Thus it is necessary for the vinegar to protect the surface of the cornichons from intrusions, but not penetrate them too deeply. In a preliminary rinsing with salt, the water inside the cornichons diffuses toward the salt. This phenomenon of osmosis produces a shriveled cornichon, which subsequently will not soak up much vinegar, especially when the vinegar is salted and sweetened.

THE GREAT QUESTION OF GREEN

Second point: How to have very green cornichons? The gathering of cooks recommends cooking in a non-tin-plated pan (until the vinegar comes to a boil for the first time). That age wanted such a pan to be copper, which reveals the reason for this point. In fact, cornichons largely owe their green color to chlorophyll molecules, which contain in their center a magnesium atom weakly bound to nitrogen atoms. When this atom is replaced by acetic acid protons, the chlorophylls become pheophytins that turn the cornichons an olive yellow. When they are cooked in nonplated copper, the magnesium is partially replaced by the copper, which gives the cornichons a strikingly green color. Now verdigris is a formidable poison. Is this practice a health threat?

As early as 1860, M.-A. Chevalier and Émile Grimaud answered yes, offering a means of recognizing if verdigris has been introduced in the preparations of cornichons in order to give them a green color: "This vicious preparation of cornichons gives rise to the production of a little verdigris (sub-acetate of copper), which can have a harmful influence on the health of those who consume it. The simplest means for recognizing this alteration consists of driving into the cornichon a needle or a *pointe de Paris*, which will be covered, after some time, with a coating of metallic copper, if the cornichon contains a small quantity of this metal."

FORBIDDEN COPPER

High schoolers, past and present, will recognize this experiment, which consists of immersing a strip of iron into a solution of copper sulfate to illustrate the so-called oxidation-reduction reactions. And toxicologists will go one step further in condemning the "greening pans" of nontinplated copper that cooks used in the past (and sometimes still today!) for making green vegetables very green. If copper is banned, how to obtain that green? For many years the food industry has been studying the addition of zinc, which also produces a lovely green. (A process called Veri-Green was patented in 1982; see "The Green of Beans" in chap. 5.)

ACTIVATING ENZYMES TO PROVIDE THE CRUNCH

Last of all, the crispness. Without special precautions, the cornichon gradually turns soft, but biochemists are studying enzymes, called pectin methylesterases, that act upon the pectins in the intercellular cement. Cornichons, like other vegetables, are composed of cells surrounded by cell walls made of three polysaccharides: cellulose, the hemicelluloses, and pectins. The pectins form an intercellular cement continuously altered by the pectin methylesterase enzymes. In effect, the pectin molecules are like long bristling threads of carboxylic acid groups ($-COOH$ or $-COO^-$, according to the acidity), of which certain ones are methylated (transformed into $-COOCH_3$). During the heating of plant tissues (in cooking, for example), softening results from a dissociation of pectins, all the more intensely as the degree of methylation is increased.

That explains the importance of the pectin methylesterases, which detach a portion of the methanol from pectins: They release carrier pectins from the carboxylic acid group. In an acid environment the pectins thus remain firmly bound. Jésus Alonso and his Madrid colleagues have shown how the skillful activation of these enzymes using moderate heat allows the cell walls to become stronger thanks to the calcium released by the cells. This treatment will allow cornichons to remain crunchy, awaiting their consumption over the course of the year.

Staleness

Water fixative emulsifiers in dough prevent staleness. Butter is not enough.

All the authors writing about childhood early in the last century, from Jules de Vallès to Renard from Marcel Proust to Pagnol, remember eating hard, stale bread, bread that could not be wasted because it was so "hard" to win. Rather than assuming today's children are still taught that lesson in frugality, the food industry is trying to prevent bread from going stale.

Heated in the presence of water, flour forms a starch; this accounts for béchamel sauce, the soft inner crumb of bread, the microstructure of biscuits. . . . This softness evolves, alas, in the storing; at different rates according to their composition, these products eventually go stale. E. Chiotelli and M. Le Meste of the University of Burgundy in Dijon have explored the interactions of starch and fats over the course of that process.

THE STORING OF ENERGY

Biscuit, cake, and bread doughs are obtained through heating flour in the presence of water. Starch granules swell and fuse together, creating a doughy system (a gel) in which the water is trapped. The molecular foundations of this phenomenon are the interactions of water and two kinds of starch polymers, amylose and amylopectin. The starch molecules are linear chains, the links of which are glucose molecules. The amylopectin molecules are branching chains, also of glucose. Insoluble in cold water, the two polymers store the glucose in the starch that is no longer leached by the water.

On the other hand, heated in the water, the starch granules gradually lose some of their amylose molecules, while the water molecules penetrate between the remaining amylopectin molecules, the structures of which are made solid by their branches. Initially wound in pairs, the amylopectin molecules unwind themselves, and the crystals that they form dissociate. That is how a soft texture is established.

Unfortunately, when the starch cools again, reassociations take place. The recrystallization of the amylopectin chains prompts a segregation: the starch on one side, the water molecules on the other. That is what is called starch retrogradation, which we can observe, for example, on the surface of a béchamel sauce kept in the refrigerator; the water beads on the surface of the congealed sauce. It is also the cause of staleness in bread and biscuits; this staleness is not a drying out, the best proof being that it is possible to soften stale bread by heating it, which restores the starch molecules to a partially starchy state in which they can recapture the water.

OBJECTIVE IDENTIFIED: WHAT MEANS TO ACHIEVE IT?

How can we combat staleness? The analysis of the mechanism leads to the idea that water must be retained and recrystallization avoided. That is why emulsifiers are used. With a hydrophilic part (soluble in water) and a hydrophobic part (insoluble), these molecules bind to the water molecules and the starch molecules that separate from one another. To that end the food industry uses lecithins, present in egg yolks or soy extracts, and various glycerides, molecules composed of glycerol, to which are bound "fatty acids." (These have a tail composed of carbon atoms C and hydrogen H, and a head formed from a carboxylic acid group –COOH.) As the water molecules are bound, the bread goes stale less quickly.

AND FATS?

Composed of a glycerol molecule to which three fatty acids are bound, the triglycerides are the major components of normal fats. Do they have an effect against staleness? And do they alter the gelatinization of the starch? Of course, the rheological behavior (the flow) of stiffened starch is different in the presence or absence of fat. Without fat, gelatinization renders the starch more viscous when the granules become so big that they densely pack the available space. The presence of a fatty substance reduces the voluminal fraction occupied by the granules, from which results the lubricating effect of the fats.

However, this behavior, disrupted or not, has no bearing on the gelatinization or the staling process. The Dijon physical chemists have shown that the triglycerides, chemically inert, form globules separate from the starch granules, and they do not make the starch granules impermeable, which would alter the gelatinization.

Nor do the triglycerides have an effect during the starch retrogradation. The triglycerides cannot form the link between starch and water, and they are too big to enter the helixes that these molecules spontaneously form, which disrupts the arrangements of the amylose.

The cook and the baker are henceforth enlightened: Butter softens dough, but other ingredients must be relied upon to produce products that retain their freshness for a long time.

Wind Crystals

Do not beat too long; prolonged beating of egg whites leads to less inflated and less resistant foam. Also, like Verlaine, add a "shiver of water over the foam."

Let us learn how to decode. When reviews publish articles with titles like "Structural and Rheological Properties of Aerated High Sugar Systems Containing Egg Albumen," we must understand it to be a matter of meringue! The *Journal of Food Science* has published the study of the British physical chemists K. Lau and E. Dickinson on the characteristics of foams composed of sugar syrup and egg white proteins according to concentrations of proteins and sugar, and beating time. The quality of the meringue depends upon these characteristics, which we examine here.

INDISPENSABLE FOR WHAT THEY ARE NOT

Foams are of interest to cooks. By providing a contrast, they allow a better perception of the texture of a dense mass in the mouth and enhance the perception of odors. (In the air bubbles the odorant molecules are more accessible.) The food industry is discovering the virtues of foams; the gas in foams reduces the quantity of matter. They can sell air!

A foam is a dispersion of gas bubbles in a liquid phase. Proteins, like those of egg whites beaten into stiff peaks, unwound by the shearing off of the whisk, are dispersed and distributed at the air-water interface. The hydrophilic parts sink into the water and their hydrophobic parts emerge into the air. Thus they reduce the "surface tension"; as the energy needed to create the interfaces diminishes, the air at the interfaces increases, as does the volume of the bubbles.

The proteins called globulins greatly reduce the surface energy. The ovomucoid and globulins retard the drainage of the liquid between the bubbles (which sustains the bubbles) because of the viscosity they impart to the egg white. The lysozyme forms complexes with the ovomucin and the other proteins, which strengthens the interfaces. In sum, the air bubbles are trapped in a rigid envelope.

E. Dickinson and K. Lau have studied elevated sugar concentrations such as those in meringues and various other airy confectioner products. The sugar increases the viscosity of the liquid, thus slowing down the drainage (stabilizing the foam) and reducing the size of the bubbles by altering the surface energy.

THE EFFECTS OF SUGAR . . . OUTSIDE THE MOUTH

To sugar syrup at a temperature of about 70°C, they added varying concentrations of powdered egg white (2, 4, 6, 8, and 10 percent) in order to obtain solutions that they then beat into stiff peaks. They measured the viscosity, the volume and size of the bubbles, and, by adding a specific colorant, revealed the composition of the protein films.

For the samples containing less protein (2 and 4 percent), the density of the foams diminished during the first 10 minutes of beating; then the expansion ceased. With higher concentrations of proteins (6, 8, and 10 percent), the density diminished during the first 5 minutes of beating, attained a minimum, and then increased again.

At the beginning of the beating, the expansion is due to large bubbles. Subsequently, the size of the bubbles diminishes as the whisk divides them without creating new bubbles. Continued beating denatures the ovalbumin, which increases the thickness of the layers at the surface of the bubbles, and the proteins reticulate there. Finally, we obtain what pastry cooks call graining; that is, the appearance of insoluble particles. Foams formed with high concentrations of proteins are more delicate, denser, and more rigid. (The films at the interface are thicker.)

The viscosity of foams also depends on the beating time: The longer the beating time, the more viscous the foam, probably because increasingly more numerous proteins are absorbed at the interfaces. For foams with 4 percent protein, the behavior is different. The shearing off coagulates the proteins into insoluble complexes, which reduces the viscosity.

With higher concentrations of proteins (4 to 10 percent), the fluidification is even clearer when the beating increases; the proteins coagulate further, leading to films more apt to rupture. Moreover, the physicists observed that the sugar recrystallized during the shearing off, the proteins probably serving as seeds for nucleation; these crystals degrade the foam by rupturing the bubble walls.

Armed with these analyses, what can the cook do? I propose "light" meringues, obtained by the addition of water to the egg white, which reduces the concentration of proteins and forms an even more delicate foam. The foam can be baked at a very low temperature for the white to coagulate before the water evaporates. That is how to obtain "wind crystals," delectable and dietary, an extremely rare combination!

That Good Old Copper

We know why egg whites beaten in copper bowls are more stable.

Does a pinch of salt in egg whites facilitate the formation of foam? Or a bit of lemon juice? Is the material of the bowl in which the egg whites are beaten significant? The adages related to egg white deserve experimental verification. At Wageningen University in the Netherlands, Erik van der Linden and his colleagues have confirmed one practice: Egg whites beaten in copper bowls are more stable.

An egg white is composed of 90 percent water and 10 percent proteins; that is, molecules formed from chains of amino acids. In the water of the egg white, the proteins, composed of hydrophobic segments alternating with hydrophilic segments, are wound into balls, with the hydrophobic zones positioning themselves at the center of the balls while the hydrophilic zones are arranged at the surface of the protein, in contact with the water.

PROTEINS AND FOAMS

The bubbles introduced with a whisk into pure water burst immediately, whereas the air bubbles formed in egg white last. In shearing off the proteins, the movement of the whisk partially unwinds them, putting their hydrophobic parts in contact with the air inside the bubbles and their hydrophilic parts into the water that surrounds the bubbles. In this way a fragile foam is formed, though durable enough for most culinary uses.

The Wageningen physical chemists measured the stability of beaten egg whites in the presence and the absence of copper ions. To do this they deposited formed foams over sintered glass funnels, collected the liquid that drained, and measured it. The results were indisputable; in the presence of copper, beaten egg whites released their water about two times more slowly than egg whites beaten without copper.

By virtue of what mechanism? Since the beating time van der Linden chose initially may not have been the one that gives foam its maximal stability, he repeated the experiment, doubling the beating speed. The difference between the beaten egg whites with and without copper was even greater. In the presence of copper, no liquid seeped out over the course of 10 minutes following the beating.

Since the stability of foams depends upon both the viscosity of the liquid phase and the surface forces, the chemists diluted the egg whites and repeated the experiments, obtaining the same stabilizing effects, which indicates that the copper acts upon the water-air interfaces rather than the liquid mass. This was the expected result, because we know that copper ions, electrically charged, can form complexes with certain egg white proteins, such as conalbumin (also called ovotransferrin because it is bound to iron ions in biological systems).

COPPER AND PROTEINS

Does the formation of complexes alter the size of bubbles in foams? Studies of foams under the microscope have shown no clear change in the structure of beaten egg whites in the presence of copper. We must look for more subtle effects, in particular, the possibility that copper ions alter the surface properties.

One of these properties is the surface tension; that is, the energy it is necessary to expend to increase the area of an interface between a liquid and a gas. If stiffened egg whites foam when beaten, it is because the proteins reduce the surface tension between the water and the air. In binding to the proteins, the copper could alter that energy. However, measurements have shown no difference for the samples with or without copper.

Other characteristics of the interfaces remain: the dilational elasticity and viscosity, which characterize the way the interfaces react to disruptions, especially when their areas are increased or reduced with the help of a vibrating ring.

The physical chemists thus discovered that, at weak frequencies (those of disruptions destructive to bubbles in the environment), the samples without copper had a modulus of surface elasticity much lower than that of the samples with copper. The bubbles with copper were more rigid and held up better. This effect is explained by the formation, already observed, of complexes between the copper ions and the conalbumin molecules distributed within a network on the surface of the air bubbles in the egg white. Thus the effect is as follows: Copper stabilizes foam by toughening the skin of the bubbles.

Italian Meringue

This is only a success if the syrup poured into the egg whites is cooked at no lower than 127 °C.

French meringue is classic. Egg whites are beaten into stiff peaks; then, while still beating them, sugar is added until the whisk turns with difficulty in the foamy mass. This mass is then deposited on a baking sheet and baked, first in a very hot oven until a superficial crust is formed, then at very gentle heat. It is a matter of partially drying out the interior without browning the exterior. Other processes also produce other meringues, such as Italian meringue, which is obtained by adding very hot syrup to stiffly beaten egg whites. At the Collège de France Chemistry Laboratory, with Raphaël Haumont, we have studied the effect of the syrup's cooking temperature on the results obtained.

When you want a topping for a cake or a tart, Italian meringue has the advantage of providing more reliable results than French meringue. In effect, when you top a tart, before baking, with beaten egg whites and sugar (French meringue), a single baking operation must bake the tart and the meringue that tops it. With just one choice of oven temperature and baking time, you will have difficulty attaining a tart baked to perfection as well as a meringue with the appropriate color and texture. Thus, the interest in Italian meringue. Pastry cooks bake the tart first and then top it with the Italian meringue, which is made independently.

Sucrose molecule

	+		Water		Water
Water	+	Water		Water	Water
Water		Water		+	Water
Water	Water	+	+	Water	
	Water		Water		+
Water			Water	+	Water

This Italian meringue brings remarkable physical and chemical phenomena into play. In practice, the egg whites are beaten into stiff peaks, and, separately, sugar is cooked with a little water. When the syrup is *à point*, it is poured over the beaten egg whites and the concoction is whisked continually until cool. The term *à point* deserves a few comments, because that is the key to success.

Let us examine the cooking of the syrup by measuring its temperature and observing the transformations that appear at the various stages of cooking. First of all, the sugar solution is transparent and its boiling resembles that of water; the temperature is thus on the order of 100°C. The liquid seems thicker when the temperature reaches 103°C to 105°C. This is the "small thread" stage. As cooking continues, the temperature continues to increase. Between 106°C and 110°C, at the "big thread" stage, the syrup, when stretched, forms a thread that reaches 5 millimeters in length before breaking. Then a surface of round bubbles forms on the syrup, between 110°C and 112°C ("small beaded" stage). Between 113°C and 115°C, bubbles will appear if one immerses a strainer in the syrup, then removes it and blows through it. The "soft ball" stage (116°C to 125°C) corresponds to the state when a drop of syrup, deposited in cold water, forms a soft ball. At the "firm ball" and "hard ball" stages (126°C to 135°C), the ball becomes increasingly harder. Then comes the "soft crack" stage at 136°C to 140°C: The drop of syrup hardens in the cold water, but sticks to the teeth. Finally, at the "hard crack" stage, between 145°C and 155°C, the drop becomes hard and brittle, without sticking. The syrup is straw colored because the sucrose is degraded and is beginning to caramelize.

Italian meringue recipes recommend pouring the syrup into the stiffly beaten egg whites, beating all the while, when the syrup is at the soft ball stage. Why? With R. Haumont, we affirmed and clarified this advice while we were exploring the physics of crunchy food matter. To obtain crunchy matter, we poured the syrups cooked at various temperatures over a cooled metallic block and obtained samples for which we measured the mechanical characteristics. We were particularly interested in the transition between two states: When the syrup was brought to 127°C, the solid that was obtained by rapid cooling was soft; on the other hand, the syrups brought to temperatures above 127°C resulted in hard, vitreous, crunchy solids.

In both cases a vitreous matter is obtained; its viscosity is so great that the molecules cannot stack up regularly as with a crystal. But for the syrups brought to higher than 127°C, the quantity of water molecules is so reduced that the interactions between sucrose molecules make themselves felt. The water, which is a plasticizer, is not sufficient to play this role and the cooled syrup is brittle.

This effect is of interest to pastry cooks. The most accomplished ones say that the syrup used for an Italian meringue must be brought to the soft ball stage. This is what our experiments indicate: When the syrup is not heated enough, it disperses well in the beaten egg whites, which it makes swell by evaporating a portion of its water, but this leads to an unstable foam, because the liquid that separates the air bubbles is not viscous enough to maintain the cohesion of the foam. On the other hand, syrup that is heated too much produces hard, brittle blocks that do not disperse well.

In other words, the analysis of the crunchiness of vitrified syrups indirectly clarifies an old procedure: To make a good Italian meringue, let us permit the temperature to attain but not exceed 127°C . . . and let us be sure to keep a close eye on the syrup as it cooks. Meringue making is an art that produces succulent results. As Pierre Dac said, "Of all the arts, the culinary arts are the ones that best nourish humanity."

The Cloudiness of Pastis

Alcohol clouds the senses; water clouds certain oils, like the anethole in pastis, as measured by neutron beams.

The idea came to him to make absinthes, and he began to add water to them, delicately, drop by drop, from time to time raising to eye level the glass in which the alcohol took on color, little by little, through the action of the water, broken down into long, cloudy spirals.

Georges Courteline, Le train de 8 heures 47

Ah, the slight cloudiness that appears when the water added to the pastis promises an imminent thirst quenching. Pastis does not have exclusive rights to this clouding. Barley water also turns cloudy, as does absinthe, with its green color that, not so long ago, took on a white tinge. Why this opaqueness? Why these changes in color? At the Laue Langevin Institute in Grenoble, Isabelle Grillo has used a spectrometer for diffusing neutrons at small angles to answer these questions in the case of pastis. The boules player's pastis is principally composed of water (55 percent in volume), ethylic alcohol (45 percent in volume), anethole (about 2 grams per liter), and various compounds that contribute to the taste of the drink. Contrary to what its name indicates to chemists (the alcohols take an "ol" ending), anethole is not an alcohol but an aromatic molecule, already used in ancient times for its therapeutic properties, to flavor foods, and to scent beauty products.

Anethole is extracted from anise and fennel seeds. It is a slightly yellow, heavily scented liquid, soluble in ethylic alcohol and not very soluble in water. Water molecules strongly attract one another and only bind weakly with anethole molecules. Thus water does not retain anethole, and the water molecules and anethole molecules constitute separate phases. On the other hand, alcohol molecules are as much attracted to anethole molecules as to fellow alcohol molecules, so that anethole dissolves in alcohol. In the bottle, pastis has a high enough concentration of alcohol for its influence to dominate: The anethole remains dissolved and the drink is transparent. When water is added, the alcohol, which is soluble in water, binds to the water, forming a primarily aqueous solution, and the anethole, which finds itself within this solution, separates from it, forming droplets, which are the cause of the clouding.

THE COLOR OF EMULSIONS

Milk, cream, and mayonnaise are all emulsions; their color is due to the reflection of light on the surface of droplets dispersed in the water. Light is also reflected on the surface of the microscopic drops of anethole. Because of this the eye sees white light (if the source of the light is white) issuing from the beverage. Moreover, the anethole droplets, the size of the light wavelength and much smaller than those formed from fatty substances, diffuse the light. In effect, the light, which is a propagation of an electrical field and a magnetic field, excites the electrons of the material through which it crosses. The electrons radiate (diffuse) in all directions. In sum, the environment is cloudy because the light is no longer transmitted but totally diffused, and objects behind the solution can no longer be seen. Thus the cloudiness of the pastis increases with the amount of dilution.

To corroborate the theory, it is necessary to measure the size of the droplets. For emulsions such as mayonnaise, in which the drops are big (on the order of a 1/10 of a millimeter), observations only require an optic microscope (the large drops reflect the light). For pastis, I. Grillo had to use neutrons diffused at small angles in order to detect the anethole droplets. The diffusion of neutrons at small angles is a powerful technique that physical chemists use to observe the form and the organization of objects between the size of a few nanometers (or billionths of a meter) and a few 10-millionths of a meter. The clouding behavior of the pure compound after the addition of water and of ethylic alcohol was compared with that of the pastis sold on the market. Pastis makers recommend diluting the pastis by five times its volume. The recommended dilution was compared with twice that dilution.

The neutron diffusion revealed that the drops of anethole have a light ray close to half a micrometer in length at ambient temperatures and just after the preparation of the sample drink. The ray of a drop does not depend upon the concentration of anethole, but the concentration in the pastis test model was a little higher than in actual pastis. Over time (12 hours for this experiment), the drops of anethole grew by about one-third. These studies also tested the effect of the temperature on the size of the anethole droplets; the diameter doubles when the temperature increases from 10°C to 40°C (but who would consider a drink at that higher temperature refreshing?), which corroborates the hypothesis that a fusion of neighboring droplets (favored by thermal agitation) engenders larger and larger structures.

The sizes, nevertheless, remain sufficiently small to diffuse the light, and sedimentation will only occur definitively at the end of the emulsion . . . but the pastis will be disposed of long before that, since, at ambient temperatures, it takes more than two days for the drink to become clear again!

The Bubbles in the Fibers

The bubbles in champagne develop especially in textile fibers, which remain on the surface of glasses.

Science demonstrates nothing, because its mission—researching the mechanisms of phenomena—is not the production of theories but the refutation of "models," a kind of simplification of the truth. Paradoxically, this work of undermining leads to advances in knowledge. A textbook case is the study of champagne bubbles. It began with the understanding that the bubbles form on the surface of glasses. Then this idea was clarified, with the observation of a few bubbles on possible "acrobats," those particles that are often in suspension in the effervescent drink. Nevertheless, the issue was not to learn whether the bubbles form on the walls or on these acrobats but to determine how the bubbles form in one case or the other!

SIC ITUR AD ASTRA

In *Casseroles et éprouvettes*, we related how a Saint-Gobain Research team had studied the surface of glasses, looking for possible defects in the glass where the bubbles might form; they imagined cracks . . . like the engravings made intentionally at the bottom of glasses.

The mechanism was supposed to have been as follows: Pouring the champagne into the glasses would have left pockets of gas in the cracks and fissures. Now the pressure of this gas, equal to the atmospheric pressure, is less than the pressure of the carbon dioxide dissolved in the liquid; thus the dissolved gas would have migrated toward these pockets, which would supposedly grow, eventually forming bubbles that would detach themselves. As the detachment of these bubbles would leave some gas in the fissures, some dissolved gas would return to enrich the pockets, forming a new bubble, and so on.

In reality, the work done by the Saint-Gobain Research team showed that the surface of glass is smooth (for the sample considered) and that the bubbles seem to form on mineral particles (tartrates, carbonates) or on textile fibers. In this new model (still false—it is the old refrain), they imagined that the mineral deposits and textile fibers formed the rough spots necessary for the formation of bubbles.

The use of an ultrafast camera refuted the model. At the Reims Laboratory of Enology, Cédric Voisin, Gérard Liger-Belair, and Philippe Jeandet have shown that the growth of bubbles primarily takes place not on the fibers but within them. Textile fibers are hollow, and pouring champagne into flutes in which these fibers adhere to the walls leaves pockets in the fibers. Computer analysis of images, produced by an experimental system according to which the camera is coupled with a microscope, has revealed that the gas probably diffuses through the walls of the hollow fibers.

THE (ESSENTIAL) FIBERS

These fibers must be considered as groups of microfibrils, where the gas dissolves in the diffuse liquid. It then comes to enrich the bubbles that remain wedged in the fibers, in such a way that the pockets of gas within the fibers (there is generally one pocket per fiber) expand and eventually "overflow" the fibers. Thus one bubble is detached, leaving a pocket of gas in the fiber, which can once again expand and engender a bubble. All that in some 5 milliseconds!

How does the bubble detach itself from the pocket of gas remaining in the fiber? The theory is not complete, but the hypothesis supposes that the Rayleigh effect (named after the English physicist) comes into play, according to which an interface, such as the one that separates the champagne from the gas, is minimized. This is the same effect, in reverse, that dissociates a uniform sheath of dew deposited on a spider web thread in the early morning into a succession of droplets: The total water to air surface is less when the droplets are formed. Here the surface is reduced when the bubble is formed.

Detached, the bubble eventually rises toward the surface. Be sure to observe this the next time you are lucky enough to enjoy the beverage attributed to the monk Pierre Pérignon: You will see that the movement is not vertical. Actually, the movement of a bubble in liquid disturbs the latter, which diverts the bubble followed by the train of bubbles leaving a particular fiber. Then too, the wall of the glass also alters the ascending movement of the bubbles, which form inclining trains. The mysteries have not lessened. For example, the follow-up on the ascent of the bubbles reveals that the surface-active molecules on the surface of the bubbles (proteins, peptides, etc.) are pushed toward the base of the bubbles over the course of their ascent. Study is difficult because the analysis of a few molecules present at one interface defies the most modern means of analysis. There is a whole world in a glass of champagne.

The Color and Taste of Port Wine

The pigments that give grapes their color are transformed over time, imparting to port its particular bouquet.

What is the relation between the song and the plumage, between the taste and the color of things, in particular, wines? Experiments involving red coloration of white bordeaux wines has shown that we are influenced by the color of foods. Nevertheless, sight is not the last word in taste: Odorant or sapid molecules can also be colored. In the case of port, there is even more: Certain molecules that confer color to the liquid are transformed, engendering odorant molecules that contribute to the bouquet.

HOW TO GIVE IT COLOR

School students cannot escape the experiment that consists of crushing spinach leaves, depositing a drop of the extract at the base of a piece of filter paper, then dampening the bottom of the filter paper with a solvent, such as naphtha. In this chromatography, the naphtha rises through capillarity into the filter paper, carrying with it the pigments. The pigments, retained differently by the paper as a function of their chemical composition, are carried along to a greater or lesser degree and form separate spots of color. Some of the spots—green—are due to chlorophyll molecules (two types of them exist in spinach), but others, in warmer tones, are due to carotenoid pigments. Among these, we can distinguish the carotenes, orange to red, and the xanthophylls (from the Greek *xanthos*, yellow). In other words, the green color of green vegetables is due not solely to chlorophylls, but to all the pigments together present in these vegetables; painters, beware!

Similarly, the color of grapes is not due solely to the molecules called anthocyanins, as was long believed. Various carotenoids abound during the formation of the grapes until the stage in which they begin to lose their green color. The concentration of carotenoids then greatly diminishes and other molecules (xanthophylls) appear.

In recent years the variations of grape composition according to pigments has been the subject of much study: according to cultivar, place of cultivation, exposure to light, state of maturity. . . . For example, the concentrations of carotenoids in grapes is higher in hot regions. This observation is important to port lovers, because the maturation of the grapes is accompanied by the degradation of the carotenoids and the appearance of odorant molecule derivatives—norisoprenoids—responsible for the bouquet typical of certain varieties of grapes.

Maria Manuela Mendes-Pinto and her colleagues at the Portuguese Catholic University at Oporto and the Avignon INRA Center have clarified the initial results by using chromatography to study the pigments of grapes used in the port wines produced in the Douro region. In their analyses, the chemists looked for pigments and also odorant compounds engendered by those pigments.

FROM CHLOROPHYLL IN WINES

An initial study showed the presence of very small quantities of carotenoids in port wines. The later study discovered eight carotenoids in a port wine aged for one year, and six chlorophyll derivatives. These molecules are missing in the wines from the Douro region. The chemists assume that differences in the production processes of these wines and of port explain the dissimilarities. Port is a "fortified wine," developed through adding alcohol over the course of the fermentation, which interrupts the latter. Ethanol facilitates the solubilization of the pigments, often insoluble in water. On the other hand, in the wines, it seems that microorganisms from the complete fermentation of the wines would degrade the carotenoids or the chlorophylls.

In young ports, the presence of carotenoids and chlorophylls or of their derivatives is an element in the final taste of the product, because these molecules are transformed into odorant molecules that determine the bouquet of the drink.

A PRECURSOR OF ODORANT COMPOUNDS

This hypothesis was tested by analyzing port wine that the experimenters aged in an accelerated fashion. The chemists compared a control port with the same port oxygenated at three temperatures (20°C, 40°C, and 60°C). These analyses showed, for example, that one xanthophyll pigment, lutein, is degraded more than beta-carotene, whatever the temperature and the oxygen content. Lutein degrades at a constant rate, and certain volatile molecules result from its transformation. They are found in old ports, whereas the young ports are richer in carotenoids and chlorophyll derivatives.

Would you like your port to have a remarkable bouquet? Now you know what you must do.

Pressure Cookers Diverted

The distillation of alcohols is regulated by the government, but not the distillation of juices, bouillons, stocks, or jams. Let us distill to retain the flavor of fruits.

The odor of jams changes over the course of cooking. Strawberries, for example, take on a "stewed" odor; black currants, an earthy odor. Cooks blame this on sugar caramelizing on the bottom of the pan . . . but at Frederiksberg University in Denmark, Camilla Varming, Mogens Andersen, and Leif Poll have shown that the changes in the odor of black currant juice are due to the avoidable loss of odorant molecules and reactions favored by the acidity and the heat.

Black currants are an important fruit. They are made into juice, syrup, gels, jams, creams. . . . In most of these productions the black currant fruit is heated and the odor changes. Any treatment of the fruit changes its taste, because it is accompanied by alterations in the concentrations of odorant molecules.

LOSS OF FRESHNESS

More than 120 odorant molecules have been identified in black currants. Listed among them are the esters, resulting from a carboxylic acid (a molecule bearing the –COOH group) binding to an alcohol (with the –OH group). These esters are generally volatile, because the association of the acid group with the alcohol group reduces their affinity for the water molecules of which the berries are mostly composed. Equally abundant are alcohols and terpenes, ubiquitous molecules in the plant kingdom.

During heating, those molecules insoluble in water and with weak molecular mass are quickly lost through evaporation. Other molecules, also insoluble in water but heavier, are carried off by the water vapor. At equilibrium, these molecules are divided between the juice and the air above it; nevertheless the water vapor formed during the heating carries off with it a share of these molecules that have passed into a gaseous phase, so that new, heavy molecules leave the solution to reestablish equilibrium between the liquid and the gas, are carried off, and so on. In the same way, concentrations of certain odorant molecules increase over the course of the heating.

C. Varming and her Danish colleagues heated the juices in closed receptacles at different temperatures. Short heating times (80°C for 4 minutes) altered the odor very little, just like pasteurization at 88°C for 27 seconds. The tasters perceived no difference between the slightly heated juices and fresh juices. On the other hand, for juices cooked at 100°C the differences were notable. Chemical analyses have corroborated these observations by showing which molecules or classes of molecules decreased or increased.

The case of esters is the simplest one; their concentrations diminished through evaporation over the course of heating. Alcohol concentrations decreased as well, but their reactivity heightened their disappearance. For the other classes, interpretations are more difficult. We know, for example, that pyrazines are formed during the thermal degradations of amino acids (black currants contain these) and that the plant pigments called carotenoids are degraded in heat, forming molecules like beta ionone (the odor of violets) or beta-damascenone. In black currants, the monoterpenes that bear alcohol groups are often combined with sugars; during heating in the acid environment of the black currants, hydrolysis releases both terpenes and sugars, and so on.

EVEN IN KITCHENS

How to deal with fruits during your next culinary manipulations? The simplest way is to use a lid; an odor in the kitchen is a symptom of odorant molecule loss (logically, kitchens should not smell good, because then we would be sure that the pleasing odors remained in the pots). Furthermore, those with access to a pressure cooker will be able to remove the safety valve and fasten onto the fitting some rubber tubing, which can be cooled by a few turns in a cold water bath, letting the water vapor and odorant molecules that would have escaped with it drip into a carafe. These molecules will form a little drop of essential oil, which deserves the gourmand's highest consideration.

Unfortunately, the distillation of vapors, the recuperation of essential oils, and their reintroduction into the juices is not enough to give cooked juice the taste of freshness; some molecules are irreversibly transformed by the heat. The Danish chemists have proposed a solution for the food industry, based on the fact that sensorial analysis has not shown perceptible differences between juices heated to 60°C for 30 to 60 minutes and pasteurized juices. Thus, their proposal is to distill under vacuum at below 60°C. To make our jams, let us try to distill by heating the fruits under reduced pressure by using, for example, a vacuum pump.

From Molecular Cuisine to Culinary Constructivism

At the end of this grand tour of the culinary realm, there are a few paths we can take to determine the future (at least the near future) of cooking. It is common knowledge that television, radio, and the newspapers continually feature the newest style, which has led from nouvelle cuisine to fusion cuisine; that is, molecular cuisine. What is molecular cuisine?

In the 1980s, when molecular gastronomy was created, its program was flawed, as I have said, because we were confusing science and its applications. In particular, we wanted to introduce into cooking new ingredients, utensils, and methods, to invent new dishes. The confusion persisted for a long time, which led to the promotion of a kind of "technological" cooking, the effective result of which was employing new jelling agents, additives, colorants, odoriferous compounds or extracts, and new materials for filtering, distilling, heating, and cooling (liquid nitrogen!).

This form of cooking is not molecular gastronomy, since molecular gastronomy is a science, but it follows from it, and journalists ended up calling it "molecular cuisine." It is a style, which means it will pass. It is not useless, because it corresponded to a moment in the history of cooking when cooks finally agreed to update their practices, to alter their ingredients, utensils, methods.

What new tendency will be next? "Note-by-note cuisine," which proposes to compose dishes molecule by molecule (or, more accurately, compound by compound) instead of using the complex mixtures that constitute classic ingredients (carrots, meat, eggs, and so on), could be developed right now, and there are already a few cooks in the world who practice this new cuisine.

Nevertheless, if the range of possibilities is immense, if the new territory to explore is gigantic, it cannot be said that this tendency is very worthwhile, because it is quasi "combinatorial," and combinatorics have never provided meaning. Monkeys hitting typewriters keys have virtually a zero probability of producing the *Odyssey*, and there would be a lot of waste.

"Culinary constructivism" seems to be a more promising trend. This is a matter of constructing dishes with a view toward obtaining predetermined gustatory effects. More essential here than rethinking traditional culinary ideas is bringing into play a conception of the whole dish, based on the laws of physiology, cultural habits, and so on. The construction must involve forms, colors, odors, tastes, temperatures . . . all aspects of the dishes . . . with a view toward producing happiness, since that is the objective that makes the work of the cook worthwhile!

The Birth of Culinary Sayings

Do sayings develop when recipes are difficult to implement and run the risk of going wrong?

Why do we hear even today, in a era when probes are being sent to Mars, that a menstruating woman will ruin the mayonnaise? Or that egg whites rise better when they are always beaten in the same direction? No doubt because cooking has developed empirically. A study on sayings, proverbs, and culinary practices seems to reveal that the recipes most in danger of turning out badly are the ones most extensively described. Can this hypothesis be tested?

The millions of French who cook perpetuate ancient practices, execute age-old recipes, propagate medieval beliefs. . . . No surprise in that: Cooking brings physical transformations and chemical reactions into play, and it is a difficult art. As early as 1742, the cook Marin wrote, "The science of cooking consists of decomposing, of making meats digestible and quintessential, of drawing out the light, nourishing juices. This species of chemical analysis is truly the whole object of our art." Of course, but how can we practice wisely this "species of chemical analysis"?

We know that chemical reactions sometimes engender dangerous com-pounds. From which arises the question: If cooking calls chemistry into play, and if chemistry presents dangers, to whom must we trust the creation of foods? To the chemists, who know the dangers of their science, but not our food culture? To the cooks, who have not mastered the science of chemical reactions? History has decided: Cooking seeks to avoid the dangers of chemistry by repeating recipes that have been put to the test. That is why cookbooks change very little; that is why, whereas a chemistry book more than ten years old is out of date, today's cooks still look to formulas that date from the Renaissance, even to Roman times.

DELICATE RECIPES, EMPIRICAL ACCUMULATION?

This conservatism explains why sayings, proverbs, techniques . . . are perpetuated, even when they seem false. It does not explain how such sayings came about. Let us examine a recipe for pear compote, taken from a 1905 cookbook: "*Take* a dozen **pears** of medium size, peel them and put them one-by-one into cold **water**. Over low heat *make* 125 grams of **sugar** pieces melt in a saucepan with a little water. As soon as the sugar is melted, add the pears, sprinkle them with lemon juice if you want the pears to remain white. If you prefer them red, you must not add the lemon juice and you must **cook** them in tin-plated copper." In this text some words are in bold type; these are the definitions. Other words are italicized; these are technically useless. The rest constitute "points of information," right or wrong, the category in which figure technique, sayings, proverbs. . . .

HYPOTHESIS REFUTED

We might assume that these points of information arise in inverse proportion to the "robustness" of the recipe, that is to say, as an inverse function of the difficulty of implementing it. That would be a plausible assertion, like the ones of which cookbooks are constituted. To verify it, let us measure this robustness. For example, for a beef roast weighing 1 kilogram, the cooking time at 180°C must be between 20 and 60 minutes. Thus, the admissible interval for the length of cooking is 40 minutes. If it were shorter than the time required for the possibilities of intervention, the recipe would run the risk of failing. In this case, the cook can relax, since taking a roast out of the oven takes less than a minute.

It is these two values we must take into account to evaluate robustness. We obtain a number without dimension (comparable to the relative robustness that might be measured using other parameters) by dividing the admissible interval for a parameter (here the cooking time) by the smallest action possible (here the duration of intervention). For the beef roast the robustness relative to the cooking time will ultimately be $(60-20)/1$, that is to say 40. The diagram shows points of information gathered from 348 French cookbooks, published between 1310 and the present, according to the robustness of various recipes. The number of directions varies, as predicted, almost inversely to the robustness . . . except for bouillon!

Bouillon stands out because, as the "soul of households," it enters into the creation of consommés, soups, stocks, and sauces, in which vegetables are cooked to give them taste. . . . It was included in all the cookbooks (and it could also be bought already prepared). It is not surprising that it has engendered so many points of information, sayings, and other mnemonic devices, which arise in two cases: When recipes are easy to spoil, and when they are vital.

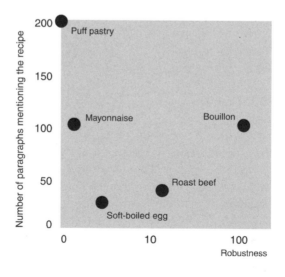

Cooking Supports Chemistry

Acidity can be measured with the help of food products. Why not take advantage of this in the laboratory? Chemistry will truly turn into cooking.

Lentils cooked in so-called hard water soften only with great difficulty, or indeed not at all. The calcium ions present in this type of water bind the pectin molecules that are in the plant cell walls, which blocks the softening process. The effect is easy to verify. Let us take two saucepans, one with distilled water, the other with water to which calcium ions have been added, and let us heat them. After half an hour of cooking, the lentils in the distilled water are puréed, whereas the other saucepan contains only small hard seeds.

Which leads us to the value of sodium bicarbonate in cooking: It makes calcium carbonate precipitate, which allows the lentils to cook despite the deleterious calcium ions, and in addition, it helps to electrically charge the pectin molecules, which, repelling one another, contribute to the softening process. Unfortunately, bicarbonate is hard to proportion correctly, and, in excess, it gives lentils a horrible soapy taste. How can we avoid this catastrophe? By neutralizing the solution with the help of an acid, which can be vinegar, lemon juice, and so on. The culinary question then becomes, "How must the vinegar be added?"

THE ACIDITY IN THE MOUTH IS NOT THE PH

Tasting is not enough, because our mouths indicate to us the acidity in the mouth and not the actual acidity, as the experiment shows that consists of first tasting vinegar (pH about 2–3, yuk!), and then the same vinegar with a large quantity of sugar added. The pH does not change, but the solution loses its acidity in the mouth.

To answer the question of neutralization, chemists, in their laboratories, have used colored indicators: phenolphthalein, bromothymol blue. . . . To make an initial link to cooking, chemists could use natural products that have indicator characteristics. For example, tea turns from brown to yellow when lemon is added (and it turns back to brown if bicarbonate is then added). The phenolic compounds in red fruits turn from red to green (with raspberries, the effect is striking) when they are alkalinized with soda . . . but then, be careful not to consume them.

And if one is color-blind? Even so, one can see fruit or vegetable slices darken when they are cut, because the polyphenol molecules in the plant tissue are altered by enzymes released only when the tissue is cut and not when the cells are intact. In other words, new molecules can appear . . . and serve as indicators that are not colored but are still visual to the extent that a color-blind person sees the shades of gray.

And if one cannot see or does not have the time to observe? Let us think about smell. At the University of North Carolina, Kerry Neppel, Maria Oliver-Hoyo, Connie Queen, and Nicole Reed have looked for systems analogous to the polyphenol systems, but odorant. Do molecules exist that are chemically altered into odorant molecules, the odors appearing only during a change in pH?

OLFACTORY TITRATIONS

These chemists established for themselves some necessary conditions: The odorant products released during an acid-base titration could not be toxic. They had to be distinct, powerful, and, if possible, of an agreeable odor. Vegetables contain an abundance of such compounds. Plants of the *Allium* genus (onion, garlic, shallot, etc.), in particular, are renowned for this. The odor of these plants appears only when the plant tissues are damaged, a condition necessary for the enzymes released by the cell damage to transform from odorless precursors into odorant molecules.

For example, alliine, an odorless molecule in garlic, is transformed by an enzyme called alliinase into a small molecule called 2-propene sulfenic acid. This molecule undergoes an autocondensation that engenders alliicine, an odorant precursor. The alliicine then dissociates into many sulfured molecules, among them a disulfur that is responsible for the odor of garlic. The reaction takes place in a neutral or acid environment . . . but not in a basic one. It is a likely candidate for olfactory titration.

The odor of garlic bothers us? Let us use the onion. The reactions are analogous, but in this case the unstable 1-propene sulfenic acid leads to an odorant precursor and to a lacrymogenous compound, S-oxide of thiopropanal.

Practical application of this knowledge is simple: Let us cut onions or garlic in a basic solution, which will prevent the odor from appearing. Then let us add a neutralizing acid. At the moment the odor appears (a small ventilator placed behind the saucepan can be used), the cook knows that the solution is neutralized.

If the alliaceous plants bother us, the range of odor molecules sensitive to acidity is vast, since the vanillin of vanilla and the eugenol of clove also undergo alterations perceived at pHs in the 6 to 8 range . . . in which case we must eat lentils with vanilla or clove!

Is It Time? You Say the Word!

Cooking uses few well-defined chemical compounds and, because of this, wastes many ingredients. What edict forbids us from changing our habits?

An apple cut and left out in the air turns brown because enzymes and polyphenol molecules released from the cells react, forming reactive quinones that engender brown compounds. To combat this phenomenon, cooks add lemon juice. The ascorbic acid in this liquid prevents browning. Would it not be more rational to use ascorbic acid directly? A molecular gastronomy success story: The lemon is supplanted in Alain Ducasse's great cookbook.

Since the way has been opened, could we not follow that lead? Can we cook with additives and colorants? Can we use aromatizing compositions?

ADDITIVES OR INGREDIENTS?

The additives first. Gelatin, wrongfully accused of carrying prions and transmitting mad cow disease, is often replaced as a texturing agent with alginates, carraghenes, agar-agar, gums . . . products formerly held in contempt because they were additives! The message is clear: A pure "chemical" product is no worse than the impure natural product.

All classes of additives are not equally useful. If texturing agents are now "cleared" by the cooks, preservatives have not (yet?) found their way into dishes made for immediate consumption. Antioxidants are sometimes used (apples, bananas, mushrooms, avocados . . . turn brown). And colorants? To color their dishes, cooks have long used saffron, cochineal, and various preparations not always well defined chemically. The food industry is more regulated, and the "chlorophyll" that it uses, for example, is much more pure than the "spinach green" prepared by cooking spinach leaves over very gentle heat, so that a green foam (the "spinach green") floats to the top of a brown juice. Cooking pasta in water colored with natural coloring agents (water turned yellow by onion skins, red by red cabbage, blue by beets) or artificial ones (apocarotenal, beta-carotene, chlorophyll, rocou) does not cause much excitement.

During an INRA seminar on molecular gastronomy, the debate raged on. It was the aromatizing compositions that sparked the most opposition. These compositions are blends of odorant molecules identified in natural products (the estragol in tarragon, the limonene in lemon, the beta-ionone in violets). Generally extracted from plant products and assembled by aromaticians, they are the food industry's version of the perfume industry's "noses."

Can a blend that smells very much like strawberries be called "strawberry aroma"? The debate goes to the heart of the cooking profession, which cannot make up its mind about what boils down to giving the taste of strawberries to strawberries, or to a preparation that does not even contain strawberries. On the other hand, certain cooks applaud the idea introduced in the "Science and Gastronomy" column of *Pour la Science*, in its April 1995 issue, which would make available solutions of these molecules, thus adding "notes" to the cook's piano. We were dreaming—we are still dreaming—of a kind of cooking that would even use specific, perfectly artificial molecules for their gustatory properties. We can transcend nature.

THE WASTE OF REDUCED WINES

While we wait, we have revisited the making of a wine sauce. Traditionally, the wine in a meurette sauce is cooked for a long time with onions, button mushrooms, and small strips of bacon. The wine in a bordelaise sauce is simmered with shallots, pepper, thyme, bay leaf, and beef bouillon. During cooking, a delicious odor escapes from the saucepan; a portion of the odorant molecules, which are usually hydrophobic, pass into the air over the saucepan, and are carried off by the water vapor.

What a waste! What remains in the saucepan? Wine is, first of all, composed of water, ethanol, glycerol, sugars (glucose, etc.), tartaric acid, polyphenols (tannins and others), and hydrophobic molecules. In the twenty-first century, why could we not produce a sauce beginning with water, glucose, tartaric acid (very "elegant," tastewise), and polyphenols, such as certain producers extract from grape seeds, for example? And keep wine for another use?

Try this test. The sauce has the red color of a red wine sauce, and an astonishing taste, which can be strengthened by that of beef bouillon cooked for a long time in the presence of glucose. In effect, the amino acids in the bouillon will have had time to react with the glucose and engender sapid and odorant compounds through reactions called the Maillard reactions (after the Nance chemist Louis-Camille Maillard), and through the many other reactions that determine the taste of foods.

Is it time? It is up to you to decide, and to say the word.

Making the Pianocktail

This machine makes new dishes à la carte.

Boris Vian once imagined a "pianocktail," which would prepare drinks appropriate to the music one was playing. With Christian Hofmann and Volker Hessel of the Institute for Microtechnology Mainz (IMM), we created the prototype of a machine that makes dishes or new cocktails, not beginning with the feeling expressed by a melody, but with a formula designed to describe dispersed systems.

In December 2002, during the Conference of the European Colloid and Interface Society in Paris, I presented a classification of complex dispersed systems that would lead to the creation of new dishes. We have already taken a quick look at this classification. Let us take another look to see how it leads to the pianocktail. It was a matter of generalizing at more than two phases the foams, emulsions, suspensions, gels, aerosols. . . . Understanding that we eat nothing but systems we now call "dispersed" was only half the battle. We had not generalized Lavoisier's idea. In a memorandum to the Academy of Sciences related to the dissolution of metals in acids, Lavoisier wrote: "In order to better show the state and to offer directly, all at once, the results of what happens during the dissolution of metals, I have constructed a specific type of formula that resembles algebra but does not have the same goal and is not derived from the same principles; we are far from that time when chemistry will have the precision of mathematics, and I invite you to consider these formulas as notations with the object of facilitating the operations of the mind."

BETTER ESPOUSED, BETTER COOKED

We would like to delve deeper into dispersed systems. We lack words, and to compensate for this inadequacy, we have introduced four symbols: / to indicate a dispersion, + to note a mixture, ø for a superimposition, and @ to denote an inclusion. The usable phases are the gases, liquids, and solids, which we will designate by their initials (in French), G for gas, H for oil, E for water, S for solid.

What can we make with such symbols? Formulas! For example, an emulsion of the oil in water type, such as cream, will be designated by the formula H/E. Whipped cream is $(G + H)/E$. The whisk introduces air bubbles, leading us to describe the physical chemical reaction in the way of chemists: $H/E + G \rightarrow (G + H)/E$. Formulation leads naturally to invention. Beginning with the three phases, combined with the help of the four symbols mentioned earlier, a thousand formulas can be envisioned, and about 50 million with six phases.

In January 2003, I proposed to Pierre Gagnaire that we create a dish called Faraday de Saint-Jacques, corresponding to the formula $((G + H + S_1)/E)/S_2$: The gas is air; the first solid S_1 is the flesh of scallops. The oil H is an oil in which orange peel has been steeped, and the water E is a smoked tea; the solid S_2 is obtained through the jelling of gelatin.

SERIES AND PARALLELS

That which is systematic lends itself to automation, so we proposed to the engineers at the IMM research facility to put their microreactors in parallel and in series to make material products from the formulas. These microreactors are large matchbox-like devices, made of metal plates into which channels have been microengineered. By assembling two plates, a double entry collector results, into which can be injected water, oil, gas, solids (powdered). Such microreactors form emulsions, foams, suspensions . . . with the proportions of the two phases determined by the outflow of the pump that pumps the phases into the channels.

Thus in February 2003, as a test run, we used puréed potatoes, oil, and water, and put in series two microreactors to materialize the formula $(G + H + S)/E$. How did this new dish taste? It was bland, because the purée lacked salt, water has no taste, and sunflower oil does not achieve gastronomic heights. On the other hand, the prototype demonstrated that an infinite number of new creations, dishes or cocktails, are within the reach of cooks and the food industry. The issue for cooking henceforth is no longer the structure of dishes, but their taste.

Terra Incognita

We have only discovered a tiny part of the culinary world. Caution, nevertheless: Exploration is not without gustatory risks.

Hic sunt leones: There are lions here. That warning appears on old maps beyond the frontier between the known, the Old World, and the unknown. The traveler was thus informed of possible dangers. Today the development of note-by-note cooking has cooks confronting the same situation.

Note-by-note cooking? That is the temptation of the chemist who, as researcher of mechanisms, trades in his scientific cap for the molecular technologist's, that inventor who applies the mechanisms discovered by science. One example: The chemist explores the production of balsamic vinegars from Modena (see Plessi et al.). He discovers their composition . . . and he is obviously tempted to reproduce the costly liquid through simple means. Sacrilege? The Italians were the first to study this question, because of the difficulty of producing the costly vinegar.

NOTE-BY-NOTE CUISINE

Actually, this famous vinegar is obtained through the concentration of a grape must from (principally) the Trebbiano vine, over the course of slow heating, until it is reduced to one-third of the original volume. Yeasts then develop in this environment rich in energetic molecules. They transform the fructose and glucose that are found in acetic acid. Then the product is stored in wooden casks that become smaller and smaller as the liquid evaporates. It takes many years of patience to produce true balsamic vinegar, which explains its high cost.

A certain industry does not hesitate to sell vinegar with a little caramel added under the same name; a dishonest enterprise, because the labeling is intentionally misleading. But when we do not have the means to acquire the true traditional product, could we not put scientific findings to good use to create an interesting substitute?

Concentration? It is obviously a temptation to use rapid means (reverse osmosis, and so on). Storage in casks? Why not be content with soaking wood shavings? The Trebbiano vine? Could we not use any white or purple grape, or apple juice? And then, sin of sins, could we not avoid the fermentation stage and use additives to directly produce a solution rich in sugars, a little acid, with strong density, concentrated in the molecules that have been detected there?

ALONG THE WAY

Let us step back a moment before setting out—or not—to discover the vast culinary world. Traditional cooking uses meat, fish, eggs, vegetables, fruits . . . which are more or less fixed molecular structures. For example, the carrot (or more precisely the carrot root) is a cellular construction with a structure principally composed of cellulose, hemicelluloses, pectins; in its cells are sugars (glucose, fructose, sucrose), amino acids, organic acids, carotenoid pigments, various odorant molecules. . . . In music, we would say this is a chord. And classic cooking, which packs into its saucepans carrots, wine, turnips, meat, is only a kind of music by chords.

WHAT NOUVELLE CUISINE?

The culinary world, which is nothing but artifice in the etymological sense of the word (have we ever seen a stew in nature?), has already crossed the frontier of this gustatory world of chords. When the carrot in the saucepan lacks sugar, the cook adds some, just as the cook adds salt when that is lacking. Here we are approaching new territory, territory in which the cook mixes classic ingredients and isolated "notes," to more subtly tune the results.

Notes? Cooking uses salt and sugar, a little ethanol, but not as yet the various amino acids (except for glutamic acid, in oriental cooking), or the various organic acids, or. . . . The list of possibilities is huge. It has been explored a bit by the food industry, with the products called additives, which frighten consumers with their coded names E330, E405, E110. . . . Nevertheless, either these products are "good," and ought to be sold to cooks, or they are "bad," and ought to be prohibited. Some are indispensable, like ascorbic acid, or vitamin C, which works better than lemon juice to protect cut vegetables from the browning caused by enzymes.

Will we enter this new world mixing chords and isolated notes? We can just discern, far off in the gustatory future, an all note-by-note cuisine, whereby the cook would forget about carrots, turnips, and so on, to bring together in the saucepan molecules in known quantities in order to obtain specific effects. I am not advising any cooks to set off into that unknown territory, where there may be lions . . . but I know it is too late. Humanity set off to explore the New World; it will create marvelous balsamic vinegars, but not from Modena.

Culinary Kinships

By classifying recipes according to their different microscopic characteristics, we make them easier to understand and stimulate creativity.

For the cook, béarnaise sauce and crème anglaise belong to different realms. The first is savory; the second, sweet. In cookbooks, béarnaise sauce is not listed in the same section as mayonnaise. One is hot and the other cold. Nevertheless, these preparations and many others (hollandaise sauce, aioli, pastry cream, etc.) are characterized by related rheological properties (the flow in the mouth), and all contain egg. Would the use of this ingredient in a sauce be a criterion for bringing together apparently unrelated preparations? Yes, the physical chemistry of dispersed systems does seem preferable and, furthermore, would facilitate the teaching of culinary techniques. For epistemologist Émile Meyerson, science was not the search for causes, which are too complex and go back too far, but the search for laws or explanatory mechanisms, because those order the world.

CODIFICATIONS, SIMPLIFICATIONS

Cooking needs these simplifications. The *Guide culinaire* edited by Auguste Escoffier, for example, offers more than five thousand recipes. What a maze! For the sauces mentioned above, physical chemistry indicates that the egg proteins play two roles. First, unwound by the shearing off that accompanies the beating of the sauce, they arrange their hydrophobic parts in contact with microscopic droplets of fat, and their hydrophilic parts, immersed in water, stabilize the fat droplets into an "emulsion." Second, the heated proteins form microscopic coagulated aggregates that are dispersed in the water of the sauce. The latter is thus a "solid suspension."

In dispersed systems such as this, composed of a continuous phase and a dispersed phase, each phase is a gas, a liquid, or a solid, which leads to new types of simple systems (see diagram).

Can this classification be used to describe the whole of cooking? Nearly, because the majority of dishes are dispersed systems: Meat, fish, vegetables, and fruits are composed of a liquid phase (intracellular fluids) dispersed in

	Dispersed phase		
	Gas	Liquid	Solid
Gas	Gas	Aerosol	Solid aerosol
Liquid	Liquid foam	Emulsion	Solid suspension
Solid	Solid foam	Gel	Solid suspension

a solid matrix (solidified by cell membranes and sometimes by cell walls). Sauces are essentially dispersed systems of the emulsion, suspension, or gel type. Rare are the foods that are not dispersed systems; otherwise, chewing would be painful. (Consider chewing on an ice cube or a large sugar crystal!)

Obviously, the new categories of simple dispersed systems are insufficient, because dishes are often the object of elaborate manipulation. Thus milk is an emulsion of droplets of fatty matter in an aqueous solution, but the continuous phase also contains casein micelles, aggregates of proteins "cemented" by calcium phosphate. When milk settles, the fatty matter comes to float on the surface, and we obtain cream, a concentrated emulsion that is churned.

As systems of more than two phases are legion, we will consider dispersed phases in the various basic phases. For example, a purée is analyzed roughly as an ensemble of cells dispersed in an emulsion, due to the fat of the milk and melted butter dispersed in the water of the milk. The cells are themselves analogous to aqueous vesicles in which starch granules swollen by the cooking of potatoes have engendered a gel.

This classification thus obtained, the taste artist will be able to make variations. Adding basil or eliminating thyme will not change the basic physical chemistry of the dish, and the interminable list of sauces (more than three hundred in the *Guide culinaire*) will thus be structured. In this way all the procedures that apply to a dish in one category can be used for all the dishes in that same category. For example, meat, fish, and vegetables having the same physical-chemical structure can thus undergo the same types of preparations, such as frying, braising, browning, grilling . . . and certain ingredients in emulsions can be replaced by others.

The description of cooking in terms of dispersed systems allows for the invention of new dishes. For example, would we like to invent a new vegetable preparation? Cooked in a bit of water, then finely chopped, the vegetables will first produce a suspension. If part of this suspension is crushed in the presence of olive oil, an emulsion will be obtained in which the proteins or the phospholipids of the cellular membranes will serve as surfactants, stabilizing the oil droplets. Provided that the emulsion has been whipped, it will contain the bubbles that are the exclusive right of foams. The mixture of two colloids will lead to a complex colloid that can be turned into a mass by gelification. . . . In the end, what will we call this dish and what will it taste like? As for the taste, it will be whatever we have decided upon. It will depend on the base vegetable and the seasonings used. Beginning with carrots, for example, the name could be "emulsified foamy gel of carrot suspension," unless in honor of the great physicist Irving Langmuir (1881–1957), who was a pioneer of dispersed systems, we were to adopt a name like "carrots à la Langmuir."

Let Us Construct Dishes . . .

By playing with combinations we will obtain associations of structures and tastes that prolong the pleasure experienced in the mouth. We will construct dishes.

If the contents of a dish constitute only a mouthful, the pleasure of eating will be slight. The German writers Wolfgang von Goethe and Friedrich Schiller identified in epic poetry what they thought to be an essential quality, the "delaying motifs." We see the outcome, but reaching it will take a long time. And in cooking? The most rapidly assimilable food substance is the liquid. This is the zero degree on the "epic" scale of food consumption. At the other end, the pure solid is the infinite degree, which cannot finally be chewed.

And between the two? The well-mannered prefer the foams, because they avoid the "disgraceful" movement of chewing. The foams are dispersions of air bubbles in a liquid, belonging physically to the emulsions (dispersions of one liquid in another liquid not initially miscible), to the gels (dispersions of a liquid into a continuous solid phase), and to suspensions. All these dispersed systems are at the one degree of alimentary delay.

NONPERIODIC STRUCTURING

To get a glimpse of the other degrees, let us draw inspiration from the natural world. We eat meat and fish, vegetables and fruits with a variety of consistencies. From what do they derive? In all these cases, there is a liquid, or a gel, an emulsion, a foam in a more firm envelope. For fish or meat, the envelope is that of the muscle fibers, bound by connective tissue, made of collagen (a protein with molecules that unwind into triple helixes that are juxtaposed in a nonweave). For fruits and vegetables, it is the same thing, but with a variety of sizes and shapes for the "cells" (in plants, the collagen is replaced by the cell wall, made of cellulose and pectins).

Since Plato claimed that art imitates nature, let us transpose these natural systems. If we arrange cannellonis vertically into a pile, and then pour into and between them a liquid with gustatory interest (bouillon, fish stock, etc.) in which gelatin has been dissolved, we obtain a mass of fibers adhered by the gel and filled with that same gel. Between the teeth, these "fibers" will have the firmness of the cannellonis, and then the fluidity of the gel.

The structuring of alimentary space is a dramatic way of constructing dishes. From which arises a new question: how to structure the space? There is nothing like a good system. One adaptation consists of first designating the components in the dishes by their dimensions: from zero (D_0), for a small grain of something that is eaten, to three (D_3), for a volume. Certain dishes are obtained through combining these elements. For example, if the notation $\emptyset_z$ is used to describe a superimposition according to a vertical z axis, the formula $D_0\emptyset_z D_3$ will describe . . . a cherry on a cake.

Of more interest is puff pastry, obtained by the successive folding of an envelope of dough (made from flour and water) into which a layer of butter is first enclosed; 6 successive folds into 3s leads to a total of 730 layers of dough $(D_{2,1})$, alternating with 729 layers of butter $(D_{2,2})$. From which arises the formula for puff pastry: $(D_{2,1}\emptyset z D_{2,1})\emptyset z_{/2y}$. Since cooking is not limited to the stacking of objects, at least two other operators $\emptyset x$, $\emptyset y$ are necessary to describe arrangements in the two other directions of space.

DESCRIPTION AND INVENTION

Let us abandon descriptive play for inventive play: With the symbols D_0, D_1, D_2, D_3 and the operators $\emptyset x$, $\emptyset y$, $\emptyset z$, we will engender formulas from which we will make dishes. That is, $((D_{3,1}\emptyset_x D_{3,2})k\emptyset y(D_{3,1}\emptyset x D_{3,2})k)l\emptyset z((D_{3,1}\emptyset_x D_{3,2})k\emptyset y(D_{3,1}\emptyset x D_{3,2})k)l$. The P $(D_{3,1}\emptyset_x D_{3,2})k$ part is an alternating line of k elements $D_{3,1}$ and $D_{3,2}$. Let us arrange l lines in the (x, y) plane by moving them forward one unit volume: We obtain the bidimensional checkerboard $((D_{3,1}\emptyset_x D_{3,2})k\emptyset y (D_{3,1}\emptyset x D_{3,2})k)l$ . . . which we can fold again.

What does this dish taste like? Whatever we desire. In a recipe made by Pierre Gagnaire, based on our proposal, French fries and strips of cuttlefish fat are prepared. They are dipped in beaten egg and alternated to compose a cube, the lateral face of which is a checkerboard. Then the cube is cut into slices, perpendicular to the axis of the strips, and one slice out of two is moved forward. In this way, a three-dimensional checkerboard is obtained that must then be heated in order to coagulate the egg, which will hold the whole thing together. The texture of this generalization of the thousand layers will be novel, and the taste will be one of the elements used . . . which is easily modified. For example, the strips could be dipped in powdered spices before assembling them.

For other constructions we can take advantage of the system of crystallography, based on symmetries. Everything benefits the cook whose calculating vision excites the imagination!

Abstract Cuisine

Our imaginations interpret the works of artists to incite an emotion. Since music and painting have evolved toward abstract forms, let us consider a more conceptual cuisine that also moves us.

Nonrepresentational painting has its theoretician, Wassily Kandinsky, who proposed replacing the imitation of nature with painting "states of the soul disguised as natural forms." A marvelous project, which the culinary arts have not yet embraced. Let us pose the question: What would an abstract, nonfigurative cuisine be? It requires the disappearance of traditional products like carrots, turnips, tomatoes, meat, fish; their "gustatory form" would be identifiable. By form, we mean "recognizable structure": The arrangement of the stars in the Big Dipper is a visual form. The arrangement of notes in *Au clair de la lune* is an auditory form; absurdist thought, an intellectual form.

AT LAST, THE NONREPRESENTATIONAL!

Molecules of cellulose, pectin, sugar, carotenoids, and so on, make up a "gustatory form" for carrot when these molecules are organized into the form of a carrot. Were we to use only the molecules that make up the taste of a carrot, we would have still made a representational dish. For abstract cuisine, the project is to use molecules to create novel gustatory forms.

Would this enterprise be the doom of agriculture and livestock production? No. Just as the abstract painter buys pigments, the nonrepresentational cook would use molecules, perhaps those of the tomato or the carrot, but not organized in the form of a tomato or carrot. Cooks would begin with isolated fractions of vegetable or animal products, which would permit the lifeblood of France to produce novel groupings of these fractions and sell them at high prices. Why would the food industry not expand the "breaking down" of milk and flour, already in operation for decades, to grapes, meat, fish (surimis are a product of such an enterprise)? These products would be assembled, but contrary to what the culinary critic Curnonsky advocated, they would "give to things the taste of what they are not."

THE ABSTRACT, IN PRACTICE

Let us now define our goal: producing a dish that would not be recognizable as a known food product, nor as an assembly of such products. The taste, the architecture, the color of the dish would be novel. Let us thus observe that nature produces no fruit, vegetable, meat, or fish in the shape of a pyramid. In choosing this visual form, we will avoid comparison to other classic foods. Let us continue. Since foods are mostly composed of water, we will make a gel in which the water is structured. Novel possibilities are numerous, because there are so many protein and polysaccharide jelling agents. And to avoid lapsing into classic gustatory forms like cheese, aspic, cooked egg, to avoid a flaked structure such as we find in the flesh of turbot, for example, let us organize this gel in the form of a three-dimensional checkerboard, alternating between hard and soft textures.

The hard cubes will be obtained from precipitating the proteins from ground beef (for example, a cut from the front of the beef will allow us to do trial runs at limited cost, since this cut will only be used for its proteins), with the help of salt, then by redissolution and coagulation by heating. The soft cubes will be obtained by grinding the same meat, but the ground meat will have a liquid added to it, before being coagulated by heat, as in the production of a mousseline.

The color? Nature has produced all the colors of the spectrum, with the help of chlorophylls (green), carotenoids (red, orange, and yellow), anthocyanes (from red to blue), and betalains (the pigments in beets). We will have trouble producing a color that is not recognized as the color of a natural product. What is left to us is black, absent . . . except in the ink of cuttlefish, and white, which is obtained by dispersing microstructures (air bubbles, oil droplets) into a phase. The three-dimensional checkerboard will alternate white and black.

And finally, the taste? We must compose the taste, the odor, the trigeminal sensation; that is, the piquancy and the coolness. For the taste, I propose a combination of many kinds of molecules. First of all, I propose the glycyrrhizic acid of licorice, but a small enough amount for the licorice taste to occur only in a subliminal way. To this we will add lactose and malic acid, never combined in natural products. For the odor, we will operate like perfume makers, who know how to create Chanel No. 5 and other compositions that have nothing natural about them. The base molecules will be odorant plant molecules, but the whole must avoid reproducing known odorant forms . . . and "project light into the depths of the human heart."

A Last Bite for the Road

A book drawing to its close always marks a painful separation for the author, when he likes the friends who are reading it, and a sadness for readers, when they like the book they are finishing. How can we avoid this separation?

At least for this subject of cooking, there is the possibility of discussing the thousands of "points of information" collected in cookbooks (in actual fact, more than twenty-five thousand). This information is our culinary culture, amassed by cooks of yesteryear, packed into cookbooks. There is information of all kinds, regarding meat, fish, vegetables, fruits, hors d'oeuvres, desserts. . . . How can we classify it? What should we do with it? In a time of cultural transformation like our own, should we not gather these points of information quickly, in order to put them into perspective, into the museums? It has been said (but my past tense is very much a present tense) that salted water comes to a boil more slowly than unsalted water. It has been said that the salt in the cooking water of green beans makes the beans greener. That roasted meat ought to be rubbed with oil and never with a liquid containing water. That a woman's period makes mayonnaise turn (and here, my past tense forms a kind of conspiracy with a belief unfortunately still alive and well!), or the phase of the moon does that, or. . . . So many things have been said and will continue to be said as long as the mystery of transformations (that are—literally—miraculous) remain unpenetrated.

The coagulation of egg white, for example, has nothing "natural" about it. If an ice cube, a solid, melts in the heat, why does an egg white, a liquid, harden in it? Moreover, these kinds of questions are everywhere; why is the sky blue and not orange, or yellow, or green? Why? Why?

We understand that such questions can keep the dialogue going *ad aeternam*. And then too, those open questions—let us forget the color of the sky for a moment to concentrate on culinary points of information—are so many promises of answers that await experimental trials . . . not always easy to interpret.

There is also the question of tastes, which, contrary to the old saying, are always interesting to discuss. Basically, it is not the "I like" or the "I do not like" that is interesting, but the "why I like" and "why I do not like" that offer ways of mutually revealing ourselves, of gaining admission into our interlocutor's private life, and thereby creating a social bond. Why does the Alsatian like Muenster cheese and the Toulousian like cassoulet? Yes, there is the underlying culture, but what is this "culture" that we set up as a deus ex machina to explain—or rather not to explain—a fascinating phenomenon that deserves to be explored further? Could the gregariousness of our human species be the key to this question? After all, as social animals, human beings live in groups, and they are happier in groups because, biologically, forming groups is the key to their survival.

Let us ask ourselves: What is it to be happy in groups, to eat in groups? Why do ceremonies so often involve a special wine, a banquet? And why has that been true for so long, since the ancient Greeks were already sacrificing animals to the gods?

Two questions in one: the question of the pleasure of eating a particular food and the question of "conviviality." They deserve to be explored rationally, scientifically, do they not? All of us together will not be enough to resolve these issues.

There are still the questions of methods, which can be divvied out and which will make us "more intelligent tomorrow than today." A multidisciplinary approach is enriching because it corresponds to an expansion of intellectual territory and to the demise of disciplinary boundaries that may prevent us from thinking clearly.

Yes, to explore cooking, we will need the "inhuman sciences" (or, all joking aside, the "exact sciences") and the "soft sciences," which organize ideas in a nonquantitative way (the question of "how many" is not always relevant). Yes, to study cooking well, we will need history, sociology, psychoanalysis . . . alongside chemistry, physics, biology. . . . Each method (let us understand that to mean "each scientific discipline") will contribute its stone to the great cairn of knowledge that marks the marvelous world of cooking. Because here is the great question: Why is cooking such a marvelous world?

This much is clear: Separation is impossible since we have that essential question to discuss!

Glossary

Usually, glossaries are arranged in alphabetical order, but why? What if we take a different approach? Let us use the alphabetical list to locate the terms, but let us stroll down glossary lane, following the course of necessity. I propose that we begin with "molecular gastronomy," since this book is a product of that discipline, and see where it leads us.

1. MOLECULAR GASTRONOMY. The scientific discipline that studies (in particular) culinary transformations; molecular gastronomy seeks the mechanisms of these transformations. It produces only knowledge, not dishes. It is practiced in laboratories using the experimental method. Not to be confused with "molecular cuisine," which is an application.

2. SCIENCE. The activity that researches the mechanisms of phenomena; employs the experimental method.

3. EXPERIMENTAL METHOD. Introduced in its modern form in theory by Francis Bacon, in practice by Galileo, it consists first of all in the observation of a phenomenon; a phenomenon is located, which is then isolated by thought, which is already an intellectual construction. Then it is characterized (measurements, measurements, measurements), and the measured parameters are brought together to produce equations, which are solved to obtain "laws," which call for explanations, or mechanisms. Then, since we know that models extracted from reality are not reality, we know that the laws, models, or theories produced are insufficient, so much so that we draw inferences from the theories in order to set up experiments that aim at refuting (not proving or demonstrating! one only demonstrates in mathematics) the theories produced, which will allow progress. This slightly caricatured description allows us to understand that science has no

end! One real difference distinguishes science from the false sciences: Science is not self-satisfied, since it is always thinking that its theories are false. Would this be a useful criterion for separating science from the false sciences?

4. MOLECULAR CUSINE. A culinary trend ("fashion"), born of the efforts of molecular gastronomy; it consists of updating cooking materials, methods, and ingredients. Since this fashion is in fashion, it is dead from the point of view of true fashion designers. This phase of technological transfer was an indispensable transition because it consisted of avoiding the repetition that has always characterized culinary empiricism.

5. TECHNOLOGY. The utilization of science (in particular) for the perfecting of technique. This is an application of the sciences (and not "applied science," since this term is erroneous; either there is science, and thus no application, or there is technology, and thus no science).

6. GASTRONOMY. "Intelligent knowledge of whatever concerns man's nourishment" (more accurately, we ought to say "human"; let us not forget women!)" justly writes Brillat-Savarin in his *Physiology of Taste* (1825). This is not cooking, and conversely, cooking is not gastronomy. Better still, "gastronomic cooking" cannot exist, since it is either cooking (the production of dishes) or gastronomy (the production of knowledge). For those who are still of two minds about the oxymoron, I propose that we feed them exclusively on knowledge!

7. TASTE. A synthetic sensation (which is just what we have when we eat) that brings together sight, hearing, smell, "sapiction" (see no. 9), trigeminal perception (cool, pungent . . .), the perception of temperatures, and consistencies (not textures).

8. FLAVOR. A useless word in French and in English, since it only means "taste." It should be abandoned by those who introduced it into the French language.

9. SAPICTION. Since gustation is the perception of the whole taste and not just the savor (a better word than "flavor"), we lack a word to describe the perception of the savor alone. The word "sapiction" has been proposed.

10. SAVOR. Or "tastes," plural. The sensation provided by a sapid molecule after the taste bud receptors are stimulated. Formerly we believed there were four of these (salty, sweet, sour, and bitter) and we published inaccurate maps of the tongue, with the supposed divisions of the various taste buds. With great effort, we believed that umami, introduced by the Japanese, was the fifth taste . . . but why "the" fifth taste, when licorice, for example, tells us that it is not sweet, bitter, sour, salty, or umami? I believe that umami is a fantasy (see no. 35) and that sensorial physiology has shown that there are many bitters, sweets, and so on. Why not admit that the number of tastes is either indeterminate or infinite, with a number of dimensions still unknown, but equal to at least six according to good sensorial physiologists?

11. SWEET. We have finally come to believe that sweet is the taste of table sugar, or sucrose, but the various sweeteners or sugars have different tastes. Sweet therefore does not exist. There are perhaps as many sweets as there are sugars or sweeteners.

12. BITTER. Another term that suffers from idealism. A bit like sauerkraut. For those who like sauerkraut, first of all (and this could apply to cassoulet, beef stew, and so on), let us note that there are as many sauerkrauts as cooks, and that, unfortunately, not all these sauerkrauts are good; there are too sour sauerkrauts, not sour enough sauerkrauts, undercooked sauerkrauts. . . . "The" sauerkraut therefore does not exist. No more than "the" bitter, since sensorial neurophysiology has shown, by fluorescent marking of the taste bud cells, that neighboring cells do not react to the same bitter molecules. There are only bitters, plural.

13. CONSISTENCY. A property intrinsic to a food. For example, water is liquid at ambient temperature. If we execute a superb dive into water, it gently parts before us, and we arrive in the water. On the other hand, if we do a belly flop, the water molecules do not have a chance to part, because their maximal speed is the speed of sound (in water). The water thus seems to be hard, and the belly flop is accompanied by the emission of a noise equal to the pain that we feel. Nevertheless, water always has its liquid consistency. Thus there is a difference between the consistency, which depends upon the microstructure, and the "texture," which is what we perceive.

14. TEXTURE. What to add, after what has been written on the subject of consistency? That water is only one example and that all foods have a texture that differs from their consistency. For example, if we are sucking on a piece of chocolate, it has a soft texture, whereas it is hard if we are munching it.

15. TRIGEMINAL NERVE. The nerve that comes from the back of the brain and divides into three branches that innervate the nose and mouth. The receptors of this system detect pungent and cool sensations.

16. ODOR. A sensation caused by molecules called odorant molecules because they bind themselves to olfactory receptors.

17. AROMA. An agreeable odor from a plant, called "aromatic." This is not a composition or an extract that is added to a food to give it an odor. Not to be confused with "bouquet," which is the odor of wine, or with "perfumes," which are products composed with a view to giving agreeable odors to the body or to rooms.

18. CONCENTRATION. The operation of gathering in a single site a group of objects; a measure of the degree of gathering. Contrary to what was long claimed by cooking, there is no concentration during the roasting process, because the juices

escape and the meat contracts. (The meat can be considered as a liquid; liquids are not compressible.)

19. EXPANSION. Increase of volume. Cooking meat in water is not an "expansion" of the meat, contrary to what cooking long claimed, since meat contracts when it is heated (because of the collagenic tissue).

20. MEAT. Most of the time, muscle tissue. It is made of long, adjacent, aligned tubes. These "tubes" are called "muscle fibers." They are responsible for muscle contraction because when the brain gives them orders in the form of electrical stimulations, the proteins within the fibers slide against one another, which prompts the shortening of the fibers.

21. MUSCLE FIBER. Muscle fibers are composed of a "skin" and an interior. The skin is the "collagenic tissue," which is created by the entangling of collagen fibers, assemblies of proteins. The fibers contain a mass of different kinds of molecules (with, of course, a considerable number of molecules of each kind), but especially water and protein molecules, of which two are essential for muscle contraction: actin and myosin.

22. COOKING. I repent! First of all, this is not chemistry, since it is a practice and not a science. Next, it cannot be reduced to a technical activity, since the best of soufflés, well-risen, will be worthless if it is thrown at the heads of dinner guests. Cooking is first of all the love, the art, and then the technique.

23. ART. A human activity, its nature having evolved over the ages, but, since I have devoted an entire treatise to this subject, I prefer to propose here, more succinctly, to oppose it to craft. The house painter and Rembrandt are equally estimable (what merits esteem if it is merited? is it not the individual rather than the profession?), but their projects differ. In one case, the goal is to protect the wall from bad weather, in the other to move the viewer.

24. TECHNIQUE. Basically, to say that cooking is first of all the love and the art is to implicitly condemn technique, to scorn it. Nevertheless, nothing will get done without technique, and the greatest artists are also the most extraordinary technicians. The Greek *techne* means "to make," but that is also the art. The relationship between art and technique is ambiguous, and our analytical impulses must not lead to black-and-white analyses. And again, why scorn whatever may contribute to the happiness of dinner guests? Especially since culinary technique makes apparent a thousand fascinating phenomena, offered as so many questions for molecular gastronomy to explore.

25. LOVE. Let us use this name for the component of the culinary act that consists of giving happiness to dinner guests. In reality, "love" of others is necessary to make them happy, is it not? That being the case, let us hasten to add that this "love"

can be interpreted in scientific terms. Each gesture of the cook, of the maître d',
can be analyzed with a view to understanding the relationships between these
gestures and the feelings of happiness they may engender. Inscribed on the German mathematician David Hilbert's tomb are these words: *Wir müssen wissen, wir
werden wissen* (we must know, we will know).

26. CHEMISTRY. What a wonderful science! This is the pursuit of mechanisms of molecular transformation. I believe it is to our advantage to distinguish it from its
applications, which could, for example, be called "molecular technology" (even
though, in certain cases, it is not a question of "molecules," strictly speaking).

27. WATER. The great forgotten element in the tables of food composition, since we
eat, essentially, water (we are ourselves composed essentially of water). "Structured" water, nevertheless: Meat is water, of course, but this "water" (let us interpret: aqueous solution) is held by proteins, among them collagen. The proteins
"hold" water? Yes, let us think of egg white, which is made of proteins and water;
here we can see very well that the water of the egg white is "held" by the proteins.
Water is held in vegetables as well, this time in the nonfibrous cells.

28. EGG WHITE. A product that is liquid and . . . yellow! Yes, with eggs, we call something that is yellow "the white" and something that is orange "the yolk," or "the
yellow." How to make sense of this, knowing that *Le Viandier*, a fourteenth-centu-
ry French cookbook, called the egg white the *aubun* (from the Latin *alba*, which
means "white"), and the yellow the *moyeu*. How to make sense of this, knowing that cookbooks from the nineteenth century noted that egg yolks are green
when the hens eat beetles in the spring. How to make sense of this when, in an
effort at "scientification," the chemist Antoine François de Fourcroy proposed
(in the *Encyclopédie méthodique*) replacing the term "egg white" with "albumen,"
which contains the same etymological error? How do we expect our children, or
even ourselves, to find the way? We should call the egg white yellow and the egg
yolk orange, should we not? Oh, I forgot that it is often useful to consider that
the egg white is made of 90 percent water and 10 percent proteins.

29. FOURCROY. Antoine François de Fourcroy (1755–1809) was the youngest of the
chemists connected to Lavoisier, and he was highly effective in implementing
the adoption of a rational nomenclature for Lavoisier's "new chemistry." A professor of chemistry at Le Jardin du Roi beginning in 1784, he was a member of
the Royal Academy of the Sciences. He obtained barium, discovered magnesium phosphate, perfected a process for separating copper from tin, and studied albumin. He discovered "vegetable" albumin in plants, which constituted a
revolution, because it was then understood that animals (man, among them) and
plants had molecules in common, contrary to what the Bible indicated. In reality,

the "albumins" in question were what we now call, and have been calling since the beginning of the twentieth century, proteins, because if it is true that there are resemblances between collagen and ovalbumin (one of the proteins in egg white), it is also true that ovalbumin coagulates and collagen does not.

30. ALBUMINS. Albumin no longer exists, and has not for at least a century, and it ought to be time to abandon this outdated term, to speak either of proteins (when meat cooks, certain proteins coagulate) or albumins, plural, because that is how we designate a class of small, globular proteins (as for collagen, it is not a globular protein). In the egg, moreover, there is no "albumin," but proteins of many kinds: ovalbumin (this is thus one albumin), ovotransferrin, ovoglobulins, lysozyme, and so on. In meat, beef or human, there is an albumin called seric albumin.

31. CHLOROPHYLLS. The paragraph devoted to albumins, plural, inevitably calls for this one, because cooking mentions chlorophyll, whereas chemists have known for a long time that chlorophyll does not exist. There are chlorophylls, again plural, distinguished by letters: chlorophyll *a*, chlorophyll *a'*, chlorophyll *b*.... Each pigment absorbs the light differently and contributes to a particular color. Of course these various molecules have chemical characteristics in common, in particular a shared tetrapyrrole group, but this group is also shared with hemoglobin, which gives blood its red color; in hemoglobin, the magnesium at the center of the chlorophyll molecules is replaced by iron.

32. EGG YOLK. An orange product that is 50 percent water; it also contains phospholipids and proteins.

33. PROTEINS. Since we mention them so often, we should stop to consider their chemical structure. Let us think of chains, the links of which are amino acid residues. They can be stretched out or folded up.

34. AMINO ACIDS. When they react, they become bound into chains called proteins. These are also molecules that have tastes. For example, glutamic acid has a taste perceived differently by various individuals according to their genetic inheritance. For some, it is simply salty; for others, it is bittersweet; to a third group, it has the taste of chicken bouillon; and for some, it has no flavor. The same is true for the various other amino acids, all of which are very interesting. For example, the collagen in meat is hydrolyzed during the making of bouillon, engendering amino acids that account for much of the bouillon's taste.

35. UMAMI. We now return to umami, to oppose the idea of establishing it as a "fifth taste." In Japan, the umami taste is the taste of dashi bouillons, obtained through the infusion of kombu seaweed. During the infusion, two amino acids are extracted, alanine and glutamic acid, which have particular tastes. In other words,

the umami taste is not a basic taste, since it is—logically—composed of two tastes. The taste of glutamic acid is not the umami taste; it is the taste of glutamic acid . . . which is also different from the taste of sodium monoglutamate.

36. SODIUM MONOGLUTAMATE. Also known as monosodium glutamate or MSG, it is called a "taste enhancer," but we must object to this term, which is also applied to salt. Salt (see my book *Kitchen Mysteries*) minimizes bitter and heightens sweet tastes; thus it is not a "taste enhancer." Then too, we have seen that taste is the complete, synthetic sensation; poor confused consumers that we are, how are we ever going to find our way in this jungle of food products? To come back to sodium monoglutamate, it is not glutamic acid, since it included the sodium ion, which contributes especially to the taste of the molecule.

37. HYDROLYSIS. To understand this process, a little etymology is enough. It requires water (hydro-) and the reaction leads to a dissociation (-lysis). For example, the sugar called sucrose is hydrolysized in an acid environment to form glucose and fructose.

38. GLUCOSE. The molecule that serves as the fuel for our cells, because energy is stored there. At this point, it is not a lack of energy that requires us to bring this glossary to a provisional close but the observation that, by proceeding thus, from word to word, we could go on to infinity. That would be an easy escape and would not offer us the key to the mystery of life: Why molecular structures such as those of cells, with membranes made of phospholipids, with worker proteins or living bricks, with DNA or DNA responsible for genetic continuity, why are they "living"? In other words, why is meat a matter for chemistry, when muscles are a subject for biology? Chemistry has succeeded in producing ex novo a flu virus. When might it (when will it?) produce a living cell? Beginning from cooking, look how serious we become, Brillat-Savarin would have pointed out, when we cross paths with a friendly science!

ALPHABETICAL LIST OF GLOSSARY TERMS (NUMBER IN GLOSSARY)

Bibliography

WORKS BY HERVÉ THIS

À table! (Peut-on encore bien manger?). Ed. Pascal Delannoy and Bertrand Hervieu. Collective work published on the occasion of the Exposition à table (Palais de la découverte). Forcalquier: Éditions de l'Aube, 2003.

Casseroles et éprouvettes. A collection of articles from the "Science et gastronomie" column of the journal *Pour la Science*. Paris: Éditions Pour la Science/Belin, 2002.

Chocolats et friandises. The best recipes of the Académie française du Chocolat et de la Confiserie. Luzern: Éditions Dormonval, 2001.

Construison un repas. Paris: Éditions Odile Jacob, 2007.

Côté cuisine/côté labo, CNDP. CD format. Catalog to accompany scientific films, 2002.

Kitchen Mysteries: Revealing the Science of Cooking. Trans. Jody Gladding. New York: Columbia University Press, 2007.

La casserole des enfants. How to carry out scientific experiments in the kitchen. Paris: Éditions Belin, 1997.

La cuisine, c'est de l'amour, de l'art, de la technique. Paris: Éditions Odile Jacob, 2006.

"La cuisine du passé au crible de la physico-chimie un atout pour l'enseignement." Reflections on the teaching of cooking. In *Actes du Colloque de l'IEHA "Histoire de l'alimentation, quels enjeux pour l'éducation?"* 71–89. Educagri Éditions, 2004.

"La cuisine, invitation aux sciences." How to use the appeal of cooking to bring children to the sciences. In *La Main à la pâte, bilan de deux ans de rélexions*, 119–27. Paris: Éditions Delagrave, 1998.

"Le banquet du Centenaire de la Société scientifique d'hygiène alimentaire des applications de gastronomie moléculaire." In *La Société scientifique d'hygiène alimentaire: Cent ans d'histoire au service de l'alimentation, 1904–2004*, 179–201. A collective work. 2006.

Les secrets de la casserole. Molecular gastronomy through questions and short answers. Paris: Éditions Belin, 1993.

Lettres gourmandes. An art book that offers reflections on the relationship between science and cooking. Éditions Jane Otmezguine, 2002.

Molecular Gastronomy: Exploring the Science of Flavor. Trans. M. B. DeBevoise. New York: Columbia University Press, 2006.

"Matière grasse en cuisine problème central de gastronomie moléculaire." In *Lipides et corps gras alimentaires*, ed. Jean Graille, 189–230. A collective scientific and technological book. Paris: Éditions Lavoisier Tec et Doc, 2003.

Quand la science dit, c'est bizarre. . . . Collective work. Éditions du Pommier, 2003.

Révelations gastronomique. Recipes decoded by science. Paris: Éditions Belin, 1995.

Traité élémentaire de cuisine. A document that gives the foundations for enlightened cooking. Paris: Éditions Belin, 2002.

"We Eat Only Disperse Systems: The Preparation of Dishes Is Largely Based on the Control of the Microstructure of Food. . . ." In *Amyloid and Amyloidisis*, conference records ed. Gilles Grateau, Robert A. Kyle, and Martha Skinner, 510–12. Boca Raton, Fla.: CRC Press, 2005.

JOURNAL ARTICLES BY HERVÉ THIS

"À chaque enfant son goût (Les rendez-vous du goût)." *Pédiatrie pratique* no. 146 (March 2003).

"The Cooking Chemist." With Nicolas Kurti. *Chemical Intelligencer* (January 1995).

"Cooking in Schools, Cooking in Universities." *Comprehensive Reviews in Food Science and Food Safety* 5, no. 3 (2006).

"Can a Cooked Egg White Be 'Uncooked'"? *Chemical Intelligencer* (October 1996): 51.

"Comment la modélisation des recettes de cuisine peut conduire à l'allègement." With Robert Meric, Rachel Edwards-Stuart, and Anne Cazor. *Revue de nutrition pratique* no. 17 (March 2004): 78–85.

"Cuisine et émulsions." *Revue générale des routes (RGRA)* no. 809 (September 2002): 59–65.

"De la gastronomie moléculaire (résultats récents)." Proceedings from the Forty-sixth Journées nationales de diététique et nutrition (May 10, 2005): 153–63.

"De l'esprit de systèmes dans l'art culinaire." *Actualités RTE*, Groupe *Elf* (October 1998): 7.

"Faisons des expériences simples." *La culture scientifique*, Atala no. 4 (March 2001).

"Food for Tomorrow? How the Scientific Discipline of Molecular Gastronomy Could Change the Way We Eat." *EMBO Reports* 7, vol. 11 (2006): 1062–66 (doi:10.1038/sj.embor.7400850).

"Froid, magnétisme et cuisine: Nicholas Kurti (1908–1998, membre d'honneur de la SFP)." *Bulletin de la Société française de physique* no. 119 (May 1999): 24–25.

"From Chocolate Béarnaise to 'Chocolate Chantilly.'" *Chemical Intelligencer* (July 1997): 52–57.

"Justus Liebig et les extraits de viandes." With Georges Bram. *Sciences des aliments* vol. 23 (2003): 577–87.

"La création d'un annuaire de la chimie des aliments et du goût." *L'Actualité chimique* (February 2003).

"La cuisson usages, tradition et science." In *La cuisson des aliments*, Seventh Scientific and Technological Meetings of the Food Industries. *Agoral* 94 (October 5–6, 1994): 13–21.

"La gastronomie moléculaire." *L'Acualité chimique* (June–July 1995): 42–46.

"La gastronomie moléculaire." *Sciences des aliments* 23, no. 2 (2003): 187–98.

"La gastronomie moléculaire et l'avancement de l'art culinaire." *Sciences* no. 98-3 (July 1998).

"La gastronomie moléculaire et physique." In *La Science des denrées alimentaires*, Food Science, 7–11, ed. Jacques Aghion, CSIPWIC, Commissariat général aux Relations internationales de la communauté française de Belgique (CGRI). Liege, 1996.

"La gastronmie moléculaire: La chimie n'oublie pas le citoyen qui cuisine." *L'Actualité chimique* (November 2000): 58–60.

"La gastronomie moléculaire un science de l'art culinaire." *Sociologie Santé* no. 19 (July 1999): 48–62.

"La gélatine face aux extraits et aux bouillons de viande." With G. Bram and Cl. Viel. *L'Actualité chimique* (November 2000): 50–54.

"Lavoisier and Meat Stock." With Robert Méric and Anne Cazor. *Comptes Rendus Chimie* (2006) (doi:10:1016/j.crci.2006.07.002).

"Le goût, les tours de main et la science." *Le retour de la saveur*, published by the INA-PG Students' Circle. Agro Paris Grignon (May 1995): 39–44.

"Les Ateliers expérimentaux du goût, nouveauté pédagogique." *Grand N* no. 70 (2002): 63–79.

"Les chimistes nous feront-ils manger des tablettes nutritives?" *AMIPS Info*, no. 72: 64–73.

"Les chimistes nous feront-ils manger des tablettes nutritives?" Proceedings from the Robert Debré Foundation, Brussels, April 2005.

"Les livres de cuisine anciens à l'épreuve du nouveau savoir culinaire." *Revue critique* 60, no. 685–86 (June–July 2004): 546–59.

"Les rendez-vous du goût." *Pédiatrie pratique* no. 143 (December 2002).

"Les rendez-vous du goût." *Pédiatrie pratique* no. 144 (January 2003).

"Les rendez-vous du goût." *Pédiatrie pratique* no. 145 (February 2003).

"L'huile d'olive et la gastronomie." *OCL* 6, no. 1 (January 1999): 95–99.

"Liebig et la cuisson de la viande une remise à jour d'idées anciennes." With Georges Bram. *C. R. Acad. Sci.* Series IIc. Paris (November 1998): 675–80.

"Lipides et goût." *Oléagineux, corps gras, lipides (OCL)* 6, no. 4 (July–August 1999): 330–35.

"Modelisation of Dishes and Exploration of Culinary 'Precisions': The Two Issues of Molecular Gastronomy." *British Journal of Nutrition* vol. 93, supp. 1 (April 2005).

"Molecular Gastronomy." *Angewandte Chemie* (intl. ed. in English) 41, no. 1 (2002): 83–88.

"Molecular Gastronomy." *Nature Materials* 4, no. 1 (January 2005): 5–8.

"Molecular Gastronomy." *World of Food Ingredients* (April–May 2004): 22–35.

"Molecular Gastronomy: A Scientific Look at Cooking." *Life Sciences in Transition*, vol. 2, a special issue of the *Journal of Molecular Biology.* Forthcoming.

"Molecular Gastronomy. Part 2: The Paradox of Culinary Innovation." *World of Food Ingredients* (June–July 2004): 34–39.

"Pourquoi la cuisine n'est pas une science?" *Sciences des aliments* 26, no. 3 (2006): 201–10.

"Préceptes magiques, cuisine empirique." *Manger Magique*, special issue of the review *Autrement*, C. Fischler, ed. (March 1996): 136–39.

"Sauce Chemistry." *World of Food Ingredients* (September 2005): 42–44.

"Science et gastronomie." *Actes des Journées scientifiques du 2ᵉ Forum des innovations techologiques du laboratoire*, Forum Labo, CNIT La Défense (April 1996): 2–4.

"Soufflés, Choux Pastry Puffs, Quenelles and Popovers." With Nicholas Kurti. *Chemical Intelligencer* (January 1995): 54–57.

"Sucrose, Glucose, and Fructose Extraction in Aqueous Carrot Root Extracts Prepared at Different Temperatures by Means of Direct NMR Measurements." *Journal of Agricultural and Food Chemistry* no. 54 (2006): 4681–86 (10.1021/jf060144i).

"Sur la température." *Journal international des sciences de la vigne et du vin*, special issue (July 1999): 99–102.

"Vive la chimie, en particulier, et la connaissance en général." With Francine Pellaud. *L'Actualité chimique* no. 280–81 (November–December 2004): 44–48.

"Who's Who in Food Chemistry." *L'Actualité chimique* (February 2003): 62–63.

WORKS BY OTHER AUTHORS

Danièle, Alexandre-Bidon. *Une archéologie du goût*. Paris: Picard, 2005.

Barham, P. *The Science of Cooking*. Berlin: Springer, 2001.

Belitz, M., and M. Grosch. *Food Chemistry*. Berlin: Springer, 1999.

Corriher, S. *Cookwise: The Secrets of Cooking Revealed*. New York: Morrow, 1997.

Gardiner, A., and S. Wilson, with the Exploratorium. *The Inquisitive Cook*. New York: Holt, 1998.

Kurti, N., and G. Kurti, eds. *But the Crackling Is Superb*. Philadelphia: A. Hilger, 1988.

Houyuan, Lu, et al. "Culinary Archaeology: Millet Noodles in Late Neolithic China." *Nature* 437 (October 13, 2005): 967–68.

McGee, H. *On Food and Cooking: The Science and Lore of the Kitchen*. New York: Scribner, 2004.

McGee, H. *The Curious Cook: More Kitchen Science and Lore*. San Francisco: North Point Press, 1990.

McKenny, D. D., W. M. Neuhausser, D. Julius. "Identification of a Cold Receptor Reveals a General Role for TRP Channels in Thermosensation." *Nature* 417 (March 7, 2002): 52–58.

Plessi, Maria, Davide Bertelli, and Francesca Miglietta. "Extraction and Identification by GC-MS of Phenolic Acids in Traditional Balsamic Vinegar from Modena." *Journal of Food Composition and Analysis* 19, no. 1 (February 2006): 49–54.

Seelig, T. *The Epicurean Laboratory: Exploring the Science of Cooking*. New York: W. H. Freeman, 1991.

Viestad, A. *Hvordan Koke Vann*. Oslo: Cappelen, 2005.

Wolke, R. L. *What Einstein Told His Cook: Kitchen Science Explained*. New York: Norton, 2002.

MOLECULAR GASTRONOMY WEBSITES

A general site at INRA, with numerous links and documents (often nearly up to date but still under construction): www.inra.fr/la_science_et_vous/apprendre_ experimenter/gastronomie_moleculaire.

La Fondation Science and Culture Alimentaire (Académie des science). A foundation to nurture research in all areas involving culinary practice: www.academie-sciences.fr/fondations/generalities.htm; www.academie-sciences.fr/fondations/ FSCA.htm.

A perspective on anticipated developments: www.blackwell-synergy.com/doi/ full/10.1111/j.1541-4337.2006.00003.x.

A free monthly meeting of the INRA conferences on molecular gastronomy to explore culinary practices. Inscription on the distribution list at herve.this@paris.inra.fr. Accessible records on the Société française de chimi website: www.sfc.fr/.

INRA course on molecular gastronomy at INA P-G. Free public courses on the work in progress. Wegsheider Organization at Clemence.wegscheider@inapg.fr.

The business of culinary invention with Pierre Gagnaire. A site where, each month, an application of molecular gastronomy proposed by Hervé This is "put to the cooking test" by Pierre Gagnaire. Free access to the "invention" and to the recipes that use it. www.pierre-gagnaire.com/francais/cdmodernite.htm.

Experimental workshops in taste. Cultural activity sessions involving cooking for schools and so on: http://crdp.ac-paris.fr/index.htm?url=d_arts-culture/gout-intro.htm.

Molecular gastronomy clubs. A new method of scientific communication put to use for transmitting results in molecular gastronomy. L'Institut des hautes études de goût, de la gastronomie et des arts de la table, an organization for high-level training, for listeners from all countries: www.iheggat.com/.

Online conferences: http://w3appli.u-strasbg.fr/canalc2/video.asp?idEvenement=312; www.canalu.fr/canalu/chainev2/utls/vHtml/o/programme/63/canalu/affich/.

Index